D0723385

In My Family Tree

In My Family Tree

A Life with Chimpanzees

Sheila Siddle

with Doug Cress

Grove Press
New York

PHOTO CREDITS: Chapters 1, 3, 4, 5, 6, 7, 8, 9, 10, 11, and 13 by David Siddle; Chapter 2 courtesy of Siddle family; Chapter 12 by Ingrid Regnell; Chapter 14 by Steve Robinson. Photo section—page 1: Clive & Tim (Siddle family); Keith & Sheila (Siddle family); Sheila & car (Siddle family); Sheila courtesy (Siddle family); page 2: Pal & David (Siddle family); Pal (David Siddle); page 3: Boo Boo (David Siddle); Donna (David Siddle); Chiquito (David Siddle); page 4: Philippe & Sheila (David Siddle); Sandy (David Siddle); page 5: Sheila & Philippe (David Siddle); Patrick & Jane Goodall (Steve Robinson); page 6: Noel & Choco (Steve Robinson); Dave & Stephan (Siddle family); Bush walk (Steve Robinson); page 7: Billy (David Siddle); Collapsed wall (David Siddle); page 8: Pippa (David Siddle); Sheena (David Siddle); Doc (David Siddle).

Published simultaneously in Canada
Printed in the United States of America

FIRST GROVE PRESS PAPERBACK EDITION

Library of Congress Cataloging-in-Publication Data

Siddle, Sheila, 1931–
In my family tree : a life with chimpanzees / Sheila Siddle with Doug Cress.—1st ed.
p. cm.
ISBN 0-8021-4010-6 (pbk.)
1. Chimpanzees—Zambia—Anecdotes. 2. Wildlife rescue—Zambia—Anecdotes. 3. Chimfunshi Wildlife Orphanage. 4. Siddle, Sheila, 1931– I. Cress, Doug. II. Title.
QL737.P96 S54 2002
599.885'096894—dc21 2001058637

Design by Laura Hammond Hough

Grove Press
841 Broadway
New York, NY 10003

03 04 05 06 07 10 9 8 7 6 5 4 3 2 1

To all those we couldn't save . . .

Contents

Foreword

by Jane Goodall

Sheila Siddle's *In My Family Tree* tells a story that needs to be told. It is an improbable tale: A farmer and his wife, knowing nothing of the behavior or requirements of our closest living relatives in the primate world, get caught in a spiraling series of events at their farm in Zambia that lead to the emergence of one of the largest sanctuaries for orphan chimpanzees in the world.

My own involvement with chimpanzees started when, at twenty-four years old and straight from a secretarial job in England, I was given the opportunity to try and find out about their lives in the wild. Two years later, when the necessary seed money and permissions had been given, I began the long-term, ongoing study of the Gombe chimpanzees in Tanzania. During the years since 1960, I have been privileged to learn the secrets of their lives in the forest and eventually—with a team of field staff, students, and scientists—record all aspects of their ecology and social behavior.

Sheila and David Siddle's first involvement with a chimpanzee was when a desperately wounded and sick infant, whose mother had been shot, was brought to them to be tended at the Chimfunshi homestead. Pal was the first in a series of hurting, bewildered, and traumatized young chimps—seized as they were being smuggled into Zambia from the great forests nearby and destined for the international entertainment and medical research industries.

News of the Siddles' extraordinarily successful enterprise soon spread far and wide in the primate conservation community. So when I met an aging female chimp named Milla, living a lonely life in a cage outside a bar in Arusha, Tanzania, whose owners begged me to help them find her a good place to go, I immediately thought of Chimfunshi. Would the Siddles take a fully grown, home-raised female? Of course, they said (not realizing quite how old Milla was!).

And so I first met the legendary Siddles when I arrived with a large and definitely disgruntled Milla on a small single-engine aircraft late one evening. We'd been flying since dawn, with one stop for immigration and customs, and we still had a drive of several hours. Milla had been tranquilized for her transference into her traveling cage. During the flight, she had woken up. Twice, when she became frustrated by her cramped quarters, she was given a small top-up of the tranquilizer by the veterinarian traveling with us, but she was very much awake when she arrived.

What was she thinking during that journey? Born in the rain forests of Cameroon, she was orphaned when her mother was shot. A British couple took pity on the infant, offered for sale next to the butchered body of her mother in the meat market. They took her to Kenya, then left her with people who would become her longtime caretakers in Arusha when they had to leave the country. Until she reached adolescence and became potentially dangerous, she had the run of a hotel and its grounds. Then she bit someone and had to be caged. On and off during the day, she would imperiously order delighted visitors to fetch her beers or Coca-Colas. It was no life for a chimp, but it was the only one she knew. She had been surrounded by friends, pampered and loved. There had been many tears when she left Arusha that morning. And now? She must have wondered what on earth was happening.

Even I, knowing how much better life was going to get for her, couldn't help crying when I crept out to see her in the moonlight, lying in the strange cage where the only things she knew were her blanket and the old tin cup she had used for years. She looked so alone, lying on a bed of straw completely covered by her old blanket—and she had no way of knowing why her world had suddenly been shattered.

The first thing Milla did the next morning was break out of her new quarters. Instead of the panic I expected, however, everyone was calm. As she had with me, Milla made instant friends with the Siddles and their head keeper at the time, Patrick Chambatu, and we all wandered around, watching as Milla raided the outside fridge, loaded up with cans of beer, tried mixing one of them with cement, used a shovel, and generally had a wonderful time. What a joy it was to find people who really understood chimps, and who took Milla on her own terms. We spent much of the day strolling through the bush with Milla, and when I left, I wrote to her anxious friends in Arusha to tell them that she was in the best possible hands.

I have been back to Chimfunshi once more since that first visit in 1990, and on both occasions, I was impressed by the way the Siddles put chimpanzees first. There are literally chimpanzees everywhere. The newest arrivals live in cages built onto the outside of the Siddles' home, so that humans can keep an eye on the orphans through glass partitions and the infants can look in on the humans, too. Your every move is watched by many pairs of curious chimp eyes. In the daytime, most of them go out with staff and volunteers, into the surrounding bush that was once grazing land for David's cattle. Meanwhile, the enclosures for the older chimps who are too big and potentially dangerous to walk the woods with human companions get larger and more ambitious.

Dave and Sheila are determined to provide the love that infant chimpanzees need so desperately for normal development, love that these chimps lost when their mothers were killed. The youngest sometimes share the Siddles' bed, and their food—from their dawn bottles to their warm bedtime drinks—is prepared in the Siddles' kitchen. And, as they grow older and spend much more time walking around outside, so they gradually regain their sense of fun and adventure. These chimpanzees who arrived as pathetic and traumatized orphans gradually reveal their own unique personalities as they gain a sense of identity and self-confidence. The Siddles and their staff know and love them all.

Once you know—really understand—chimpanzee behavior, you learn to be humble. We humans are unique, but not as different as we used to think. Dave and Sheila know how blurred the dividing line is between ourselves and our chimpanzee relatives. They know that chimpanzees are capable of complex intellectual performances, that they can be socially manipulative, that they experience happiness and sadness, anger and jealousy, fear and despair.

This book will bring tears to your eyes as you read some of the stories, the descriptions of the terrible condition in which many of the babies came to them. And there are stories that will make you laugh. The Siddles help many other orphaned animals as well, such as antelopes and monkeys—and just wait until you meet Billy, the all-but-adult hippo who still drinks milk from a bottle and only stopped lying on the living room sofa when it collapsed beneath her weight. Above all, you will be filled with admiration at the way the Siddles have jumped in to help each and every orphan. They never refuse an animal whose life has been torn apart, who needs understanding and love if he or she is to become a self-confident adult. No sacrifice has been too great. I am sure every reader will feel about the Siddles as I do, and I hope you will feel moved to reach out and help them in their never-ending struggle—physical, mental, and financial—to mend broken lives and provide a brighter future for each and every chimpanzee who comes their way.

Sheila and Dave, I salute you—for your dedication, your hard work, your persistence, and your courage. You are an inspiration to us all.

One

Pal

Pal in 1983

They say you'll never forget where you were and what you were doing at momentous occasions in your life, but I honestly can't recall the circumstances of October 18, 1983. I don't remember if it was cloudy or sunny; I don't know if I was working in the garden or feeding the geese. I'm not even sure what time it was. All I know is that I held a dying chimpanzee in my arms that day and it changed my life forever.

Pierre Fabel was an honorary game ranger who was also married to my daughter, Diana. I'd received a message around midday over the short-wave radio that he was on his way out to Chimfunshi, our farm along the Upper Kafue River in central Zambia, but nothing prepared me for what lay ahead. Pierre got out of his truck and approached, carrying this pathetic, terrified animal in his arms. It was a baby chimpanzee, but not like any chimp I'd ever seen before. What hit us first was the smell. My initial reaction was one of nausea and horror, and Dave, my husband, felt the same way. This small chimp—a bag of bones, really—had badly smashed teeth and the right side of his mouth was slit open about two inches more than it should have been. His face and his mouth stank from the rotten flesh around the badly infected wounds. He was totally dehydrated and suffering from terrible diarrhea, and there were flies everywhere around him.

Pierre said the chimp had been confiscated from some poachers who were caught smuggling him into the country from Zaire. Although officially protected as an endangered species, chimpanzees flowed illegally into Zambia in those days, and we heard of quite a few people owning them as pets. Game rangers traditionally did nothing to confiscate the animals, since there was no facility in place for keeping them, and this was quite possibly the first time anybody had done anything in response to the black-market trade. The little chimp's entire family had probably been killed for meat, and now he was headed to the market to be sold as a pet—though it seemed unlikely he'd even live that long.

Pierre was visibly upset.

"If you don't do something to help him," he said, tears welling in his eyes, "then I've got to do something about him myself."

I hate to think of what Pierre had meant by that, but I knew he'd certainly seen plenty of terrible things as a game ranger, and the sight of him being so overcome by emotion came as a shock. I looked down at this poor, sad little chimp. His mouth flopped open in a gruesome grin, exposing the left side of his jaw and gums, and his teeth looked as though they'd been hammered to bits. His eyes were dull. His breathing was labored. It was obvious he was dying, and yet here he was, clinging to life.

Dave and I knew nothing about chimpanzees. We were cattle farmers, and even though we'd spent most of our lives in Africa and encountered all sorts of wildlife, this was like nothing we'd ever seen before. But it was obvious what we had to do. Because this chimp looked so much like a human baby, we immediately began treating him like one, and luckily, we soon found that once you've got a baby chimp in your arms, instinct takes over. We carried the chimp into the house, and the first order of business was to try to clean up his wounds. He struggled a bit, and my attempts at putting any sort of salves or disinfectants on the cuts were hopeless. In retrospect, I feel a bit stupid that I did not try to stitch the mouth closed at the edge, but at the time I was more interested in trying to simply clean him up as quickly as possible—and, anyway, we were fighting for his life. We were very scared for him then, so we decided to try to feed him with a bottle of milk. To say the chimp was overjoyed is an understatement. He sucked greedily on the teat, even though more milk kept pouring out of the gash on the side of his mouth than down his throat, but it was the most life he'd shown since his arrival.

Once the chimp accepted the bottle, it was easy to put antibiotic medication into the milk and there was no trauma in treating his infections internally. Thus began a vigil that lasted the rest of that first day, as the chimp alternately drank and slept, his breath coming so fitfully at times that we feared each might be his last. Once or twice he opened his eyes and seemed to get a clear look around at his surroundings, but then exhaustion—or relief, perhaps—seemed to overcome him and he'd drift off again. Dave and I took turns holding him as he slept or preparing his bottle, and we even took him to bed with us at night. But I don't think we ever stopped to think, "Well, now what?" We were running on instinct and there was no time to try and collect our thoughts. Just like a human child, this chimp responded to anybody who offered him a little tender loving care, and that's what we meant to give him. We christened this brave little fellow Pal.

At the time, there was no way of knowing how completely our lives were about to change simply because we'd decided to help an injured chimpanzee. Before Pal arrived, Dave and I were looking forward to retirement. I was fifty-one years old and Dave was fifty-four, and our five children had long since grown up and moved away. Chimfunshi, the old fishing camp we'd bought in 1972 near the headwaters of the Kafue River and turned into a fifty-five-acre cattle ranch, was to be our final home, and we both had worked hard to make it the sort of place we'd always dreamed of. Our days were long but our lives were good, and some evenings, when we'd sit out under the big acacia trees and look west over the floodplains as the herds of antelope or elephants passed by, we told ourselves we'd found the last unspoiled place on earth.

But then Pal arrived, and suddenly everything else ceased to matter. Whatever routine and order we'd established on the farm was promptly forgotten, and our every thought centered upon the chimp and his well-being. If Pal awoke at 4 A.M., Dave and I awoke at 4 A.M. If he napped in the afternoon, Dave and I tiptoed around the house so as not to disturb him. There were warm bottles at dawn and warm baths at night. It was as if someone handed you your grandchild to raise as your own, and suddenly all the parenting skills you thought you'd stored away for good were being dusted off and put to use.

Pal weighed only fourteen pounds when he arrived, and our best guess from all that we read about infant chimps and the photos we looked at in books seemed to indicate that he was about a year old—definitely still a "baby" by anybody's standards. Yet Pal's youth seemed to contribute to his rapid recovery. Within weeks, those awful wounds on his face were almost entirely healed, and even though he would never lose the terrible scars that gave him that rather droopy look, it was not long before he seemed able to use his mouth and lips freely. We were also lucky that it was only his baby teeth he lost. Even his diarrhea began to wane. But while Pal's physical recovery was impressive, his emotional recovery proved to be a long, slow, uphill climb. His sleep was always tortured and traumatic, and his nightmares were the worst. Some nights he awoke in such a state, screaming in terror and screeching so hard that he began to shake uncontrollably. Was he remembering his capture? The death of his mother? His own injuries? I had no way of knowing, of course, but I'd hold him close and stroke his fur until he calmed down and, more often than not, he would fall back asleep in my arms.

Pal grew quickly in confidence, and soon he was venturing throughout the house, pulling open cabinets and raiding the bookshelves. But he wasn't a particularly destructive chimp, definitely not the sort to tear a house to pieces. Pal was just extremely curious. He was also terribly smart. He'd watch what you were doing, peering over your shoulder and looking intently at your face as you concentrated, then do a perfect imitation. He once spent an hour or so watching Dave cut wire with a pair of pliers, and, when Dave accidentally dropped the pliers, Pal swooped in and grabbed them, then scampered off to the nearest fence and began cutting the wire himself.

As I said, Dave and I knew nothing about raising chimpanzees. I think we had a copy of Jane Goodall's *In the Shadow of Man* around the house, but while the book offered fascinating insights into the behavior of wild chimpanzees, it said nothing about raising one in your own home. Neither did any of the textbooks or scientific journals we found, and since chimpanzees are no longer indigenous to Zambia, even locating someone who knew a *little* about chimps proved difficult. So we improvised. For instance, some animal experts say that you can't possibly potty-train a wild chimp, but Dave and I had never heard this—so we did. We potty-trained Pal in a little over four months, just by sitting him on the toilet like you would a child until he figured out what we wanted. Once he was housebroken, you'd hear him go into the toilet and then there'd be this tinkling noise. And he'd wait because he knew you'd come after him, and then you'd look into the toilet together and pull the chain to flush it, and when the water would rush in, Pal would laugh hysterically. He thought swirling water was terribly funny.

We also found ways to include Pal in our daily routine, since Dave and I both had full-time jobs running the farm and couldn't afford to spend hours watching him around the house. Zambian women wear what is called a *chitenje,* which is two meters of very colorful cloth that can be used for everything, including fashioning a sling of sorts so that one can carry a child on either one's back or front. I owned a few *chitenjes* at the time of Pal's arrival, so that was how we carried him around the farm. When I did my chores, I slung Pal up onto my back and carried him everywhere I went. And when I was unable to cope any longer, Dave took him everywhere else. Pal thoroughly enjoyed being carried around this way, and he would alternately grab at things over your shoulder and catch catnaps, when he would curl into a tiny ball of black fur. In the wild, Pal would have spent most of the day being carried around on the back of his mother, so I doubt that he felt this routine was unusual.

Pal was not the first primate at Chimfunshi, however. We already had a young male baboon at the time, whom we called Rocky because he'd been confiscated from some local boys who were trying to stone him to death with rocks. He arrived in terrible shape, too, but we gave him a lot of attention and love, and he recovered wonderfully. Rocky had a great personality. He was bitten by a snake once, and when I treated the wound, he was so proud of the bandage on his arm that he showed it off to anybody he could find. Rocky also used to ride around on the backs of my bullmastiff dogs as if they were horses.

So Pal joined our minimenagerie. I used to take them all out for walks in the nearby forests to get exercise, and we'd laugh because here I was with a chimpanzee, a baboon, and four large dogs—I know we were an odd sight. But I was pleased to see how quickly Pal reverted to chimp behavior out in the bush. He acted just like any other chimp baby would have: he played in the grass or leaves when we stopped, he climbed small trees, and he spent a great deal of time searching for his favorite fruits, like figs and cherries and *msuku,* a juicy green treat that is somewhat like a cross between an apricot and a grape. But mostly, baby chimps stay close to their mother, and since I was the mother figure, that meant he seldom let me out of his sight. In the daytime, because there were no other chimps around, Pal would often sit close to Rocky and allow him to groom him. Rocky would run his fingers through Pal's fur, removing any pieces of dirt or dry skin, and both of them seemed to get such pleasure out of the process. All monkeys and apes enjoy grooming, and Pal would sit there motionless for an hour or more, with the most faraway look on his face. I often wondered what he was thinking about at those times—happy things maybe, or sad—but the experience clearly left him more calm and relaxed.

Pal often returned the favor, grooming Rocky, and later Dave and myself as well. He sometimes got annoyed with me when he found that I had hairs growing in the wrong place on my chin or pimples that needed squeezing, and would hold my face tightly in his hands and make disapproving grunts and click his teeth together. He used his lips and teeth to pull out the offending hairs, but the pimples he would squeeze very gently between his fingers, not releasing his grip on my face until he was satisfied I looked better.

Meanwhile, Dave and I came to regard Pal just like a child. We'd already raised five human children between us, so it was relatively easy in the early stages to cope with a single baby chimpanzee, especially since he acted just like our own children had. Pal would play with toys, take naps, throw tantrums, and pout, just like any little human. He ate at our table alongside us, and drank what we drank. He loved milk and tea, and even had sips of our beer now and again. For treats, we'd slip him pieces of chocolate, which he adored, and sugarcane. You might say we went a bit too far and nearly "humanized" him—something I am careful to avoid now—but in those days, we were basically just trying whatever worked. Meanwhile, Pal became quite adept at making himself understood. About two months after his arrival, he began bringing his cup to you when he wanted a drink. He'd thrust his cup forward, and there was no doubt what he was saying: "I want a drink." And if I gave Pal a cup filled with water and he looked at it and gave it back, well, then we knew it was milk he wanted, or tea.

Pal taught us a good many lessons about chimpanzees, but nothing was more surprising than his ability to barter. He frequently got ahold of something he should not have, such as a cap or someone's glasses, and since his favorite food is bananas, I learned to offer him a banana in exchange for whatever he had stolen. If he turned his back on me, I knew my offer was ridiculous. So I would increase it to two bananas—again the back treatment; then three bananas, and so on. Eventually, as we got closer to a deal, Pal would turn and stare me straight in the eyes, holding up the purloined item, until I got to a figure he could live with—usually five or six bananas. Then he'd make a soft, pant noise that signaled his agreement, and he'd slip the goods toward me with one hand while pulling the bananas toward himself with the other.

As he grew older, Pal's ability to communicate proved uncanny, and there were times I'd swear he must have spoken to me telepathically. But nothing amazed me more than his reaction to a particularly nasty strain of flu that struck Chimfunshi once. Pal looked so pathetic that we honestly feared for his survival. For days, he lay about the house and yard, barely moving, refusing to eat any foods. Since chimpanzees do not cough or blow their noses as we humans do, they are very susceptible to a buildup of fluid in the lungs that can prove deadly. Pal looked utterly wretched one morning, with a big, hard blob of mucus hanging from his nose and his mouth slung open in a desperate attempt to breathe, and he refused both his milk and his medication. He looked at me through half-closed, bleary eyes that told me just how sick and miserable he felt, then slowly used his left hand to push away my offer of a cup of milk, while extending toward me a long piece of straw with his right hand. I was confused and offered the milk again, but Pal was adamant—he pushed the milk away while again appearing to hand me the straw. Suddenly, I realized he was trying to barter with me—but for what? I had no idea.

Pal kept repeating this gesture to me over and over, and he looked so bad and felt so hot and feverish that I was in tears.

I said to him, "Pal, my darling, I am so sorry but I do not know what you want."

At that, he put the piece of straw down and went off looking for I-knew-not-what. He rummaged through some old tires and sacks, yet returned with another handful of straw and repeated the barter gesture, but I still didn't understand and could only offer him a cuddle. Pal gently pushed me away and wearily moved off again, clearly in search of something, and then appeared to find what he wanted—a tiny piece of dried orange peel. He offered me the peel in the barter style, with a pleading look in his eyes.

"An orange?" I said. "You want an orange?"

It hadn't occurred to me that something as simple as the vitamin C in citrus fruits would help combat a chimp's cold—or that a chimp would instinctively know it—but Pal kept holding the peel up to me as if to emphasize the point, so I quickly went to the food storage hut and brought back three oranges. As I walked back, I heard this low, pleased panting, and Pal's face was positively smiling. He accepted the oranges—and gave me the peel in return—then went off to his bed to enjoy the fruit. Just like a human being, he had a craving for something. He knew he needed it, and he made that need clear to me. Over time, I came to realize that chimps—and probably all animals, for that matter—know instinctively what they need to be healthy and happy. It's we humans who get in the way and make a mess of things.

Around this time, Pal had his first fit. I heard this terrible crash early one morning and went to see what was wrong and found Pal unconscious on the floor. I shouted in shock and surprise and slowly Pal seemed to come to; then he struggled to his feet and staggered toward me with great difficulty. I pulled him up into a sitting position and found that his pulse was extremely rapid, and I could see that his head was still cloudy and his eyes were having trouble focusing. His heartbeat eventually slowed back down to normal in about fifteen minutes and Pal seemed none the worse for wear, but the incident left me badly shaken and I continued to watch him closely for the next few days. About three weeks later, it happened again, then again two weeks after that. Pal would just collapse into a heap on the ground, then lie there jerking and quivering spasmodically before passing out altogether. Yet he always came to when I shouted to him or wrapped my arms around him. Clearly something was wrong, but none of the medical experts I could find knew enough about chimpanzees to offer much help, and the best guesses were that he was suffering from a form of genetic disease such as epilepsy, or perhaps the beating he'd taken when he was captured had caused brain damage that prompted the fits. The strangest part was that Pal seemed otherwise fit and healthy and the attacks came on with no warning whatsoever—which meant that I was always on guard.

Despite the fits, Pal's strength and confidence grew, and he started to regard Dave and me as his own—so aggressively that he acted horribly toward other people. He began getting jealous if anyone came near us. If someone got too close, he would bite that person, and children were a particular target. If you think about it, most children were about Pal's size then, so I'm sure he came to regard them as a direct threat to his "family," especially since his real family had come to such a brutal end. But chimps are also four times stronger than humans—even at that young age—and the possibility that Pal might seriously injure someone was very real. Our grandchildren were regular targets, and it reached the point where they no longer felt safe when they came to visit. One day, Pal bit a young girl's foot very badly when she and her mother walked past where Dave and I were sitting, and suddenly we found ourselves with a problem chimp on our hands.

But fate intervened, on a couple of fronts. A lot of people seemed to own chimps in those days, mostly as pets, and it wasn't that strange to find a chimp tethered to a tree in the backyard of some of Zambia's finer homes. But after years of looking the other way, the Zambian government finally announced in 1984 that anybody owning illegal pets—such as chimpanzees, African gray parrots, Rosen parakeets, or other endangered species—had exactly one year to apply for a license to keep the animals, even if they'd owned them for years. I remember they ran large adverts in the newspaper and hung posters all over town, and it was the first time they'd ever put anything like that down on paper. Of course, it had been against the law for almost a decade, but now the Zambian government started to cooperate and enforce the laws established by CITES—the Convention on International Trade in Endangered Species. That's when the black-market smuggling of animals also began to come under scrutiny, and suddenly all those chimps that border guards had once not bothered to collect were being confiscated—and Chimfunshi became the animals' logical destination.

Which is why on April 20, 1984, barely six months after Pal had arrived, the game rangers confiscated another young chimp from smugglers and again Pierre came out to the farm, this time carrying a dehydrated female chimp that smelled terribly of diarrhea and was missing a toe on her left foot. This chimp was not as easy to handle as Pal, since she was older and bigger; we guessed she must have been about four or five. But even at that age she was still really only a baby herself. She would normally have been breast-feeding for her first four years, and certainly would have still been traveling and sleeping with her mother.

If Pal arrived in a sorry state, however, this chimp was positively wretched. She was so weak that she couldn't even climb onto a chair without falling, and was so timid she refused to allow anyone to pick her up when she fell. Her balance was also probably affected by the loss of the toe, which we presumed had been shot off during her capture. But we did the best we could for her. We put an automobile tire on its side on a shelf in the bathroom off the living room and filled it with straw and blankets to make a bed for her. Although she never seemed nervous about us, the only movement she made during the first few weeks was to stick her bottom over the edge of the tire to defecate—and was that a mess! There'd be blood and feces and mucus all over the bath, and the stench was a clear indicator of how ill she really was. When it came time to eat, she'd slowly sit up and nibble on some food, then slump back into the tire and lie there again for hours on end. She never objected to my cleaning her tire or feeding and grooming her, but neither did she ever show any signs of curiosity. Given this chimp's utter lack of ambition, Dave and I didn't struggle in coming up with her name: Liza Do Little.

Liza did not look like any of the chimps I'd seen in pictures. Her face was quite round, but her forehead was flat and her shoulders sat well up beyond her ears, as if she had no neck. But even though we knew nothing about her origin, the fact that she looked nothing like Pal was the first indication to me that chimps from different areas might look different, in the same way that people do. We treated her no differently than we did Pal, however, slipping antibiotics and medication into her bottles of milk in order to fight off the infections that were ravaging her internally, and working slowly to build up her trust and confidence in us.

It was probably three weeks after she arrived that Liza one day hesitantly stuck her head around the door to see what was beyond the bathroom walls, then rushed back to her tire. This began to happen more and more often, until she eventually plucked up her confidence and came and sat with us for longer periods. Then, Liza and Pal became friends. We made a place for Liza to sleep at night and slowly it became easier and easier to leave Pal alone with her, and she more or less adopted him as a little brother. But when we eventually weaned him from sleeping in our bed, both Dave and I realized how badly we missed him. After having Pal in our bed for almost six months—and after so many nights spent holding him or bottle-feeding him while he slept—we suddenly found ourselves going to bed without him and something seemed wrong. We'd wake up in the night and instinctively say, "Where is he?" He taught us a lot about chimps and just how close to humans they really are—and how much closer they can become.

Once, as Dave and I prepared to go away for a trip to West Africa, I spent a great deal of time talking to Pal. I'd say, "Right, Pal, I'm going away now for a bit. Will you be all right until I come back?" I told him that my daughter, Diana, was going to take good care of him, and I told him it was only going to be four weeks. I went over all this with Pal again and again, because I didn't know what else to do. We hadn't been out of his sight since he'd arrived, and now we were talking about being away for a solid month. I also told Pal I loved him very much. He seemed to understand and was not particularly upset as we left. And we were told he took the first four weeks' separation like a champ. But when we were delayed with plane connections in Nigeria and did not return until almost a week after we were due back, friends "greeted" us at the airport in Lusaka with the news that Pal was very, very ill. We rushed home, of course, and Diana came running out to the car and threw herself into my arms, sobbing. I obviously thought the worst and, steeling myself, asked grimly, "Where is he?"

Pal was lying on a bed, motionless. I assumed he was dead and practically fell on him, crying hysterically. Slowly, Pal opened one eye, looked at me, and turned his head away. Dave entered the room and Pal again opened an eye, looked at Dave, and gave him a very feeble greeting before once more turning his head away. Then Pal slowly sat up and weakly motioned for Diana to come pick him up. And for the next twelve hours, Pal would neither look at me nor even acknowledge my presence. Then, suddenly, he cried out and grabbed hold of me, as though meaning to never let go again. Whatever had ailed Pal seemed to have magically disappeared.

Diana told us that for four weeks he had been wonderful and really well behaved. Then, shortly after we were supposed to have returned home, he got listless, stopped eating, and could not be persuaded either to drink or to eat a thing. Diana was certain he was about to die. I've since read about orphaned chimpanzees in the wild who literally allow themselves to die of sadness after their mother is killed, withering away in a few short weeks. Pal clearly regarded Dave and me as his family, and when we stayed away too long, he apparently gave up on living as well. I really do believe he would have allowed himself to die. Common sense says that four weeks shouldn't make any difference to a chimp, yet why did he do just fine over those four weeks and then suddenly get sick when we didn't return? There was something very revealing there, and it taught me to respect a chimp's intelligence—and devotion.

Two

Africa

My father with the converted truck that carried the family across Africa

I was raised on a small farm in northern England called Moss Villa, in Lancashire, near the town of Preston. The farm was only about fifteen acres or so, but we had a very large orchard where we grew apples and pears, and I grew up with the usual sort of farm menagerie: a cow for milk, a few pigs, chickens, rabbits, ducks, geese, dogs, and cats. Our family consisted of my father, Robert Landing, whom everyone called Bob, and my mother, Matilda, whom everyone called Tillie. I was the oldest child, born in 1931, followed by my brothers Keith (1935) and Clive (1942), and our black Labrador dog, Tim. I remember being very happy there.

In 1947, Daddy and Mummy decided that a holiday with a difference was in order. My parents had always been adventurous and a little bit restless, but what they planned this time was a bit more adventurous than usual. As Mummy once said in a letter, she and Daddy "were of the same tendency, itching to be off and away to faraway places to see as much of the big, wide world whilst we were still young enough to enjoy it." They thought a trip would do us all good. So they decided that we would drive from Lancashire all the way to Cape Town, South Africa, a trip of well over six thousand miles that would take many, many months and cause us to traverse the French Alps, the Mediterranean Sea, the Atlas Mountains, the Sahara Desert, the Congo rain forests, and finally the vast grasslands of southern Africa.

What made my parents want to pick up and travel like that? I think the Second World War had a lot to do with it. My father had been a private and fought in many very nasty situations while in the Royal Army Service Corps, with my mother suffering the anxiety that went with being a soldier's wife. She refused to go into the bomb shelters during air raids, preferring instead to sit on our beds with us and speak to my father as if he were there in the room with us. Somehow, it made us all feel closer. Daddy never talked about his war experiences, but he served in France and was evacuated during the horrible retreat at Dunkirk, and I know he must have seen and felt some terrible things. I can't imagine how he suffered. The only time I can remember the war ever even being discussed at our house was when an old friend, who had been a major during the war, came to dinner and told a story about Daddy and his outfit being hungry with nothing to eat while on maneuvers in France. So my father disappeared into the countryside for hours and hours—so long, in fact, that the others got quite worried—but he returned with bacon and eggs and all the men enjoyed a proper breakfast.

I think Daddy came home from the war and found he wanted more than the ordinary out of life. Postwar England was fairly depressed, and there weren't many opportunities for a man like my father, who was something of a free spirit. Some of our relatives were concerned that my parents were doing the wrong thing in taking us out of school for such a long and difficult journey, but Mummy and Daddy were certain the trip would be a wonderful education in and of itself. My brothers and I were, of course, delighted with the prospect of missing school and traveling for months on end, and looked forward to the trip with tremendous anticipation.

My father was a cabinetmaker by trade, but he was clever and seemed to be good at most things. So Daddy sold the farm, put some money in the bank for later, then bought five old army trucks, which he stripped down and cannibalized to make one really good truck that would comfortably accommodate the entire family. The body of the vehicle had originally been an RAF wireless radio truck, but one of the trucks Daddy used had been an ambulance, so he kept the beds and we each had our own bunk in which to sleep. Even our dog, Tim, had a special bed, and the plan was for us all to sleep at night in the truck. There was a little stove in there, and places to store everything, from food to clothes. The interior of the truck was metal and it could be rather noisy, but there was a sliding glass window on each side so we could look out and another window on the back door. I loved it; everything was so new and brilliant.

It might seem odd that our dog would be included in our traveling party, but he really was a member of the family. Daddy found him when he was a nine-month-old stray, and my brothers and I adored him. Tim was a special dog and incredibly smart, and he did all sorts of clever things. For instance, we had a duck who was confined to a certain area on the farm because he was blind, and one day Tim came running up, barking his head off at me and obviously wanting me to go somewhere, so I followed him to our pond. There in the middle was the duck, who had somehow managed to find water but, being unable to see, couldn't find his way out of the pond. Tim tried several times to swim out to the duck and steer him back toward shore, but every time he got near, the duck panicked. So he fetched me instead, and I was able to wade out and rescue the duck myself.

The *Manchester Preston Guardian* newspaper heard about my family's big trip and offered to buy the rights to our story, right from Manchester all the way down to Cape Town, with the agreement that every so often Daddy would wire back updates from interesting places. The day we left, they had a picture on the front page of the newspaper of the family gathered in front of the truck, with my youngest brother, Clive, and Tim together in the middle. We said good-bye to everybody and then left. But because there were so many last-minute details and so many people to visit, we didn't actually leave until almost 11 P.M. Daddy was determined to go as far as possible south toward London that first night, however, so we set off.

The truck was built in such a way that there was a cab and a gap between the cab and the body of the truck, where Daddy built an extra petrol tank and a water tank. But between the cab and the body was a tube about twelve inches in diameter. We called it a "speaking tube," but you could actually pass things back and forth from the cab to the body. Because we were pressed for time and left late, Daddy decided to eat his dinner at the wheel, and Mummy handed his food toward the front through the speaking tube. When he was finished, Daddy had something left on his plate, and shouted, "Here, Tillie, give these scraps to Tim," and passed something back. But Mummy said, "Tim's not with us. He's up there with you." And Daddy said, "No he isn't," and hit the brakes. By this time we were very close to London, and all hell broke loose.

We had to stop the truck to examine every nook and cranny, and Tim definitely wasn't there, so Daddy said, "I'm going to carry on and get to a phone." He rung up the *Manchester Preston Guardian* from a pay phone and said that somehow or other his dog had been left behind in the excitement of departing. So the next day, the *Guardian* cropped Tim and my brother from the photo of the family the newspaper had used the day before, blew it up quite large, and put it on the front page, offering a reward for the return of the dog. And just outside Manchester, an old lady went out early the next morning to get her newspaper, picked it up, and saw this dog wandering around, one that seemed to resemble the dog in the photo. So she shouted, "Tim! Tim!" and he came running up to her.

Meanwhile, in London, we were all devastated, because we'd left Tim behind and didn't know if we'd ever see him again. My brothers and I were crying and were so upset that we refused to eat. But then we got a telegram from the *Guardian,* which said to meet the next train at London station, as Tim was traveling first-class! We all danced around and sang, then went out and treated ourselves to a proper feast. At the station, Tim actually leaped from the train right into my arms, and the photographers crowded around, taking pictures of the family reunion as the flashbulbs went off. Never again did we leave anyplace without first making sure Tim was with us.

In January 1947, we left chilly, wet England from Dover and sailed across the English Channel to Calais. We drove straight through the heart of France—which was still wet and cold—from Amiens to Marseille, passing through places like Rouen and Paris and Lyon on the way. Even though much of France was left in rubble from the war, which had only ended in Europe a year and a half before, I remember it as one long sight-seeing adventure. But some of the strangest sights were close at hand. For some reason, while we were in France we stayed one night in a hotel. That was where I saw my first bidet, which I thought was for washing your feet. I also remember the family's going to a restaurant in France and Daddy's trying to order bacon and eggs for breakfast. Daddy had a smattering of French left over from the war and he explained to the garçon that we wanted "*oeufs avec bacon*" and all that, but the waiter couldn't understand any of it. Daddy kept trying and trying, but the waiter just turned and went off. We didn't know what to expect. The garçon finally came back with plates of beautiful bacon and eggs and put them on the table and said in perfect English, "Excuse me, sir, but if you'd have asked me in English it would have been much easier."

At Marseille, we left Europe, sailing across the Mediterranean Sea on a barge to Oran on the northern coast of Algeria. From there, we climbed back into the truck and drove overland to Algiers, which was a whole new world. The heat finally hit us when we reached the Sahara Desert, where it was broiling hot during the day and freezing cold at night. But the desert was so beautiful. It undulated so wonderfully, so gracefully, with these yellow waves of sand that just rolled and rolled away. There was an awful lot of life on it, though at first you didn't realize it. So many little things—flowers and lizards and other small animals—live in and around the desert.

I was very impressed with the Arabs. They seemed to be just as I pictured them from the stories I read in school, so strong and dark and mysterious in their robes. In fact, it was near Tamanrasset that I think I first fell in love. We were going to stay there a couple of nights, and the family was just getting settled when all of a sudden galloping toward us was this white camel, and on it a man wearing a cloak and headdress. But unlike most Arabs, he had piercing blue eyes, very bright and clear, and I remember feeling as if he was looking straight through me. My heart stopped. And as he got to me he sort of reined in the camel and pulled up right in front of me. He had come to ask my mother and father if we would have a feast with him that night to celebrate our arrival and our travels, but I don't recall much of the conversation. I can't explain to you what was happening, but my heart was pounding. All I could hear was this *ka-dunk, ka-dunk, ka-dunk* sound in my chest, and all I could see were those blue eyes.

The feast was a wonderful experience, very colorful and grand, even though they sacrificed an animal in our honor, which I found rather upsetting. But nothing could shake the image of that white camel and the man on it. I sometimes wonder now if time has made his eyes bluer in my memory, or if perhaps his camel wasn't quite so white. But my fifteen-year-old heart did somersaults, and did so for many years afterward.

The Sahara Desert provided some other vivid memories. At one point, when we were about seventy-five miles outside of Tamanrasset and struggling to adapt to the heat, Daddy began to get very worried that we were going to run out of water. We had heard that Arabs often sucked on stones to keep from getting thirsty, so my brother Keith and I very dramatically began sucking on some rocks we found in order to preserve water and thereby save the family. (At least that's what we told ourselves, although we eventually made it to Tamanrasett with plenty of water to spare.) It was also in the Sahara that I almost got my head blown off—literally. Daddy kept several large guns in the front cab for protection, loaded and locked in a clip behind the seat. We children took turns sitting up in the front, in order to see more and change up the routine, and one day I was seated up front when Daddy put the truck into gear and it lurched forward, inadvertently jarring loose a double-barreled shotgun that he had not properly secured. The gun fell forward and went off with a terrific explosion right behind my head, blasting both barrels past my ear and out the open window. No one got hurt, thankfully—the shells missed me by inches—but it scared the hell out of me and *really* scared the hell out of Daddy.

It was also in the Sahara that we came across another English family making a similar trek. Sir Harold Ingrams, who had been appointed as Great Britain's Chief Commissioner of the Gold Coast, had decided to travel overland to his post, taking along several friends and members of his family. We met up with them in the little walled town of In Salah in central Algeria, and then again repeatedly for the next few weeks in places like Wadi Arak and In Gezzam and Agadès, as we wound south toward Zinder in Niger. Sir Ingrams eventually wrote a book on his journey called *Seven Across the Sahara,* in which the Landing family is mentioned here and there, and he paid us a wonderful compliment by writing, "One could not but admire the spirit which moved them to make this tremendous journey."

I don't think I had a preconceived notion of Africa. I had read a lot about it, and I had pen pals in South Africa, but when we finally reached the southern edge of the Sahara, everything seemed totally new and totally unexpected. Somehow, the colors all seemed more vibrant and the smells sharper and stronger than I could have imagined, and some of the views were so vast that it was hard to take it all in. I also remember meeting people who had never seen white people before, and the look of shock and surprise on their faces was something that will stay with me always. There were some Europeans living here and there, but being a white face that stood out was a feeling I won't ever forget.

The animals also made a tremendous impression. We saw ostriches and camels in North Africa, then little things like lizards and mice and scorpions in the middle of the desert. In British Nigeria, I got to ride Arabian stallions—big, proud animals, so regal and strong. But it was the African wildlife that fascinated me. We saw every type of game imaginable, and in large numbers, with herds of wildebeest and antelope and zebra that started at one corner of the horizon and stretched all the way to the other. And elephants—I can't recall when I first saw those, or rhinos, but one day you just realize that you're seeing them again and you're not so amazed anymore. That's how many there were back then. One day, Daddy stopped the truck as two massive herds of elephants approached one another from either side of the road. There must have been a hundred or so in each herd. We'd been told that elephants sometimes fought in situations like this, and we all craned forward to get a good look at the battle royale that would certainly ensue. But instead, each herd simply drifted through the other like ghosts, slipping quietly past one another and then heading off in their separate directions.

The only ape we saw was one gorilla, somewhere in the deepest jungles of the Congo. He was on the road, just standing there as we approached. The gorilla waited until we got quite close, then dashed over the edge into the thick forest; when we stopped the truck and looked down, he was gone. I suspect that ours was the first truck he ever saw and that he had no idea what to make of it. I doubt he'd ever seen people before, either.

When we reached the Congo, we encountered an outbreak of rabies that nearly cost Tim his life. Daddy customarily hid Tim in the back of the truck at each border crossing, since the paperwork and bureaucracy required for pets usually accomplished nothing more than to delay us for hours, and so it was at the Congo border. We were allowed through after a brief check of our papers, and, with Tim safely tucked away inside, Daddy drove on past the border gate, then pulled into a hotel a few yards down the road to fetch some hot water for tea. But when Daddy entered the lobby of the hotel, Tim followed, and the border guard just happened to look back as the black dog trotted past.

Within seconds, Daddy was confronted by the angry guard, who brandished his rifle and declared that he had no choice but to shoot Tim as a rabies risk. Daddy was amazingly calm.

"All right," he said. "But let's go behind the hotel to do it."

The two men and the dog walked quietly out of the lobby and around to the back of the building, where the guard raised his rifle and took aim at Tim. But as he did so, Daddy pulled out his pistol and pointed it at the guard's head. "Just remember," he said, "that when your gun goes off, you are dead."

The guard, clearly shaken, continued to point his gun at Tim and sputtered that all animals that did not go through quaratine would have to be shot. Daddy nodded, but repeated his threat. "If your gun goes off," he said, "you are dead."

Another second or two passed, but then the guard lowered his gun and ordered Daddy—and Tim—back onto the truck. He told us to drive off for good, and I don't believe I can ever recall Daddy going so fast.

Later in the Congo, our family was invited to dinner at a doctor's home. This doctor owned a dog who was famous for once killing a lion, and so he warned us not to let Tim out of the truck for fear of his dog's making short work of ours. But while we were trying to come up with the best way to keep Tim and this doctor's dog apart, Tim slipped out of the truck. At first he approached the other dog playfully, wagging his tail and bouncing on his toes, but when the doctor's dog began to growl, Tim flew at him. We didn't have time to interfere; within seconds, the dogs were locked in a terrific life-and-death battle. My brothers and I were screaming and I remember my mother closing her eyes and beginning to pray, but lo and behold, Tim pinned the lion-killing dog and locked his jaws around his foe's neck, forcing the doctor to sheepishly intervene and save his own dog.

When we reached Uganda in East Africa, it was the rainy season and the roads around Lake Victoria were closed. So, instead of waiting several months for the weather to change and the roads to dry out, we took a steamship across the lake, from Entebbe in Uganda to Musoma in Tanganyika on the other side. Lake Victoria seemed as big as an ocean, and there were times when you struggled to see the land. It was wonderful, because the captain made friends with us and pointed things out as we went along. That was where I saw my first crocodile, and we saw some hippos too. I also saw people in canoes, and they waved with big smiles as we chugged past their little wooden boats.

As we continued south, I could see that Africa was taking a firm hold on our family. My father and mother fell in love with what was then called Northern Rhodesia, and now Zambia. When we reached the eastern region of its forests and savannah in March, everything was fresh and green and lovely, and we all felt something change inside of us. Africa had become as much our home now as England ever was, and I think we realized that our attachment to the little farm we had left in Lancashire was slipping away. I don't think any of us missed it.

Then my mother, who was never sick, fell ill with malaria at Messina, just inside South Africa, and it was thought unwise to take her any farther. By that time, Mummy and Daddy had decided that southern Africa was the place for them, but since Daddy had contracted with the *Manchester Preston Guardian* to sell the *whole* story of our trip, he felt compelled to complete the journey. So Daddy left Mummy and us in Messina and went on to Cape Town, filed his final updates, and came back again. He then moved the family to Fort Victoria in Northern Rhodesia. We weren't there too long before Clive got tonsillitis, and when he went in for surgery at the local clinic, he unexpectedly collapsed under the anesthetic and nearly died, prompting the doctors to say there must be something seriously wrong with him. So Daddy took Clive and Mummy and sailed back to England, leaving Keith and me in boarding school in Southern Rhodesia, which is now called Zimbabwe. When they got back to England, the doctors found that Clive had simply had an overdose of anesthetic, and they took his tonsils out properly.

After Clive recovered, so did the adventurous spirit in Mummy and Daddy. Rather than sail back from England, the way they'd come, they got another truck and went overland all over again, right down that same route through Africa, until the family was back together.

When my parents had left for England, no one had known what to do about Tim, as both Keith and I were in school and could not keep a pet and Mummy and Daddy were in too much of a hurry to bother with the paperwork of bringing a dog along.

We decided to leave Tim with an elderly family named Smith in Salisbury, Southern Rhodesia. The Smiths loved him, and Mrs. Smith took him for long walks each morning and allowed him to eat her table scraps. In fact, Mrs. Smith took great pains to write my parents letters back in England hinting at how hard it would be to have to give Tim up.

My father and mother began to fear they might never get him back, but then Daddy devised a plan. He had a special way of whistling that allowed any member of the family to locate him, and which he often used when he took Tim on long walks. So when he returned, Daddy drove up to within a half-mile of the Smiths' home, got out of the truck, and said, "Right. I'm going to whistle three times and see if Tim comes." He whistled once, but there was no response. He whistled again, and still nothing. But when Daddy whistled the third time, Tim suddenly bounded over the garden hedge and raced into our arms, his tail going nonstop. Once we had Tim safely back within our grasp, we proceeded on to the Smiths' home, so that they might give Tim a proper farewell.

Three

Chimfunshi

"Chimfunshi"—Place of Water

In 1949, my family settled in Livingstone, along the Zambezi River, not far from Victoria Falls, and I finished my education at the Dominican Convent in Bulawayo. I was never very successful academically, but I scraped through and got a General Certificate of Education. Livingstone was a wonderful place to grow up. Most of our school holidays were spent cycling to Victoria Falls and playing in its mist or hiking through the forests in search of animals, of which there were plenty. We often saw antelope and zebra in the grassland nearby, and the trees were full of monkeys and parrots and snakes.

Then Daddy got himself a job as a carpenter on the copper mines in Chingola, up near the Congo border. Chingola was a boomtown then, buzzing with the activity of the mines, which were supplying a large portion of the world's copper and literally driving the nation's economy. Under the British colonial rule, Zambia was a fairly prosperous, exciting place to live, and this continued after independence in 1964. I used to visit my father at work, taking him his lunch some days when he forgot it, and never gave a second thought to climbing the long, rickety stepladder that stretched all the way to the top of the mine silos, then walking around the narrow edges to reach him.

I wanted to be a nurse, and so I worked in a hospital for three years. It was not an actual training school, but my time there would have counted toward my nursing certificate if I had gone to the training school in Bulawayo. I very much regret now not having gone for the training, but I had other things on my mind: I fell in love. In 1952, I met Don Jones, who worked as a clerk in the Government Public Works Department. We dated for a bit and then got married, and Don and I settled into Zambia's growing expatriate community. We also had three daughters: Lorraine, born in 1954, Diana (1955), and Sylvia (1957).

Government employees in those days worked for three years and then had to take a six-month holiday, which was supposedly for "climatic reasons," but was widely understood as nothing more than a means of keeping you from getting too attached to a certain place. During the colonial era, when Great Britain ruled countries all over Africa as protectorates, this policy served to instill a certain detachment from the people and the culture among workers. Over the next thirteen years, Don and I lived in Abercorn (which is now called Mbala, in northern Zambia) on the edge of Lake Tanganyika—I loved the lake and it is still one of my favorite places—and we had a three-year stint in the capital city of Lusaka, which in those days was quite small, but which for me nonetheless had too many people.

Animals played an ongoing role in our lives, and I soon developed a penchant for nursing sick and wounded creatures. Word got around, which meant we always had neighbors and friends dropping by with their pets. I don't think I had any special medical ability, I think I just got on well with animals, and maybe paid a little closer attention to their needs than other people did. For several years, we had a jackal living in the house with us, a small, doglike animal, similar to a coyote or a fox. It was a lovely jackal with a pointy snout and a soft silver coat, but it met a most unfortunate fate. We had a lady visiting once who happily allowed the jackal to lie in her lap as she nonchalantly stroked its fur. But when she inquired as to what type of dog it was, I said, "Oh no, that's not a dog. That's a jackal." The silly woman immediately panicked. She grabbed her skirt and flung the jackal off her lap so vigorously that the poor animal slammed against the corner of the wall and died of a broken back.

We also had a genet cat living with us for a time, a beautiful, spotted animal that looked like a cross between a leopard and a house cat. It belonged to our daughter Lorraine, and it really behaved like a proper pet. It used to love to attack the kids' feet when they wiggled them under the bedclothes. In 1955, we acquired a cheetah. I used to assist the local veterinarians on occasion and was helping this vet treat a pet cheetah that was suffering from rickets because its owner had given it nothing but cooked food. So I took the cheetah home and gave it the right diet—raw meat, bones, that sort of thing—and it lived in our back garden for quite a while until it grew healthy and strong enough to be returned to its owner. Years later, I nearly got another cheetah when a game ranger accidentally shot a female that had two cubs, then appeared at my door with these cubs and asked if I could help. I took one and someone else took the other, but then suddenly the government in Lusaka intervened and took the cheetah cubs away, saying they could not be kept in a private home. But neither got the proper treatment, and they both died.

In the 1950s, I came upon one of my greatest passions: motor sports. Don had raced motorcycles for many years, but I found that rally car driving was what got my heart going. The races took place all on Zambian tracks, with rough trails through the bush, and I often participated in two- and three-day rallies. We had an all-female team, and my navigator was an engineer, so she was good at math. But you *had* to be good at math, because all you were given was a sheet with directions telling you where to go, and the skill of your navigator was a big part of the equation. I drove and she did the calculations, using a slide rule, even though her head was sometimes shaking up and down too violently to read the numbers. We drove through some lovely areas, and plenty of mud and rain, too. The race wasn't always about speed; sometimes it was just about finishing in one piece. We never won a race outright, but we did finish second and third a few times. If I had won, I don't think I'd have been very popular, because it was a very male-dominated sport.

My last three years with Don were spent in northern Zambia in Ndola, another fast-growing town along the copper belt. But those weren't happy times. Don and I had drifted apart and decided that our marriage wasn't working anymore, so I applied for a divorce. Divorce was difficult in those days and was frowned upon by everyone, and it was difficult to lose so many friends. It was also terribly hard on our children, who were old enough to understand but young enough to be scared and hurt. Still, things had gone all wrong between Don and me, so in 1964 I pushed for the divorce and got it, and took custody of our daughters.

Meanwhile, David Siddle had come to Northern Rhodesia in 1954 as a quantity surveyor to work on the construction of the Kariba Dam, a massive structure that was built just outside of Siavonga near what is now the Zimbabwe border. When I met Dave in 1966, he was a building contractor with his own company, and he was overseeing the construction of a large block of offices and a bank in Chingola. Today, Dave is often described as looking like Father Christmas, with his flowing white beard and hair, but in those days he cut a rather dashing figure, and drove one of the few sports cars in the area. We hit it off almost immediately. Each Sunday afternoon we'd go for lovely drives along the dirt roads outside of town. Sometimes we'd park, too, and sit and talk about where we'd been in life, and where we wanted to go. We found we had a lot in common: Dave had also been married and divorced, and had custody of his two sons, Tony (who was born in 1955) and Charles (1959). Both of us were a bit brittle after our divorces and had no interest in getting involved again so soon after our messy breakups.

Though we thought we were going slowly, it wasn't long before Dave and I fell in love. He proposed over the telephone. I said yes, of course, and we got married on October 18, 1968. We were very fortunate to have children of similar ages, because they all got on well together and we became one big happy family.

There was never any question as to where Dave and I would make our home. Once you have lived in Africa for more than a couple of years, you fall so in love with the place that you become unable to leave. Almost from the moment my family set foot in Africa, I can't ever remember having a wish to go back to England. In fact, I've only returned a handful of times in the last fifty-odd years. Africa introduced me to people and places untouched by what we call "civilization," and it spoiled my whole family from ever being able to return to the English country lifestyle we'd left behind. My feelings are particularly strong about Zambia. Though we are surrounded by war-torn countries and have certainly felt the political and economic effects, Zambia itself has remained fairly peaceful. We've never once thought seriously of leaving. My brother Keith has lived in Zimbabwe for the last forty-five years and has seen a lot of violence and strife, but has said time and again that he could never leave his "home" either. My brother Clive is equally at home in South Africa. We may all be British expatriates, but inside, we're Africans.

Early in our marriage, Dave and I set about looking for a place in which to build a home, somewhere we could retire together. I was thirty-eight at the time we got married, and Dave was forty-one, and it was likely that this would be the last place we'd ever live, so we wanted to make it special. I remember Dave's saying it had to have a big flat tree, like an acacia, which is so beautiful and majestic and typically African, and we both wanted a place near water. We looked at a couple of sites that were interesting, but nothing that caught our fancy, until, in 1972, we heard about this little fishing camp that was for sale along the Upper Kafue River, the waterway that snakes along Zambia's border with the Congo before plunging straight down the middle of the country. Dave and I went to have a look. It was about forty miles west of Chingola along the dirt road to Solwezi, but the site itself was another ten miles inland, through thick forests and gentle streams.

When we finally emerged at the camp, it was our dream place. There wasn't much there at the time, just a couple of small shacks on a bluff at a bend in the river, and the entire property was only fifty-five acres. But the forest came right up to the edge of the water and the views were spectacular. To the east there were dense trees and vegetation that rose up like a wall, and when you looked out to the west, all you saw was river and wide-open spaces. The site was a floodplain, which meant that the grassland was green half of the year and under water the other half, but it was beautiful. There was even a big, flat acacia tree right on top of the bluff, just like Dave wanted. We asked somebody what the area was called, and he said, “Chimfunshi,” which roughly translates to “place of water” in the local Bemba language.

We bought the little fishing camp on the river, and paid just $2,000 for it—a real bargain, even at the time. Dave put his contractor’s expertise to work and built a small brick house on the bluff where the camp had once been, complete with a lovely verandah with a thatched roof that overlooked the river. The house itself wasn’t too large—just a kitchen, a lounge, and a few bedrooms and bathrooms—but it was big enough, and we planted a grove of banana and orange trees just beyond the verandah in a clearing on the hilltop. Next to that we planted a large vegetable garden, and there were soon chickens and ducks and geese and dogs underfoot. Because of Chimfunshi’s remoteness, we expected to get lonely, but we had only one weekend to ourselves in the first year. Friends and relatives kept coming out to the Kafue River and, like us, fell in love with the place.

At Chimfunshi, amenities like electricity and telephone service were out of the question, given how far we were from Chingola, and solar power was still too new and expensive to be practical. But I always preferred the simple life anyway, even if it meant doing without appliances that we had taken for granted. Instead of a refrigerator, for instance, we used a “charcoal box” to keep our meats and perishables fresh. This was a large wooden crate lined with charcoal on the outside and tightly bound with chicken wire. During the day, you poured water on the charcoal so that it stayed cool and kept the inside of the box at about 50 degrees Fahrenheit, even though we stored it outdoors. I made my own cheese and butter, and we often bartered those items for goods and services when we made our weekly trips into Chingola. Later, we purchased a gas-powered generator that gave us a few hours of electricity.

In 1974, the Zambian government began allocating land for farming in our region and Dave was offered a vacant farm and ten thousand acres of grazing rights adjacent to our fifty-five-acre plot. Although we had tried chicken farming for a few years after we got married, both of us wanted something a little more ambitious, so we bought a hundred head of cattle and became cattle farmers. Zambia was trying to encourage its beef industry in those days, and the grassland and climate of central Zambia is very good for cattle. Our herd eventually grew to over a thousand head, and we later diversified the herd by importing Brahman bulls into the country. Dave spent most of his days checking the cattle and managing a staff of herdsmen, and we eventually added a dairy herd that gave us a consistent supply of fresh milk on the farm. I ran the business side of things and looked after the children, although they were often gone for long periods when they were away at boarding school. Eventually they either married or moved out to live on their own, until it was just Dave and I again. The work was hard and the days were long, but sometimes, when the jacaranda blossoms drifted down from the trees like purple snow, or when the lightning storms flashed along the copper veins and lit up the whole sky with terrific claps of thunder, we wondered how we could have ever lived anywhere else.

Four

Chimpanzees

The Youngsters enjoying their morning milk

One chimp around the house was manageable, and even two weren't that difficult to mind. But once Liza arrived in early 1984, word got out that Chimfunshi was taking chimps, and suddenly we began receiving orphans at an alarming pace. Two came in May 1984 and another the month after that. By August 1984 we had a total of six chimpanzees to care for, and Dave and I found ourselves struggling to keep up. It wasn't just chimps, either. We had begun to receive scores of other orphaned animals by then too, like monkeys and baboons and parrots and tortoises and antelopes and even dogs, some confiscated from poachers and others found by the local SPCA, but all in various stages of distress and all in need of a home. What amazed us was how quickly the situation changed, too. In the blink of an eye, Dave and I were building cages and nursing sick animals practically around the clock.

Yet chimpanzees remained our primary focus. When Pal arrived, I don't think we ever sat and thought, "Well, now what are we going to do with this one chimp?" simply because he was still a baby and we didn't know enough about chimps to realize he would one day be so strong and smart and quick that the human mind would reel. But when we got a few more chimps . . . well, that was when we suddenly realized that we were not chimp keepers, that we had no facilities, no experience. What were we going to do?

Looking back, I'm amazed at what we didn't know. We had Jane Goodall's book, which was helpful at times, but I honestly can't remember how we learned. Or how we coped. Simply treating each chimp like a human baby was fine when they were alone or in very small numbers, but trying to communicate with six young chimps was a little like trying to bring an unruly kindergarten class to order. Just when you thought you had all their attention, one would hit the other or pinch this one or slap that one, and suddenly the whole mob would dissolve into a roly-poly ball of arms and legs and fur. The sheer numbers also required at least three people to be on hand at feeding times, since they all had to be given their bottles simultaneously or another great fight would break out. Believe me—chimps *do* cry over spilled milk.

The first additions to arrive were Girly and Junior, a pair of two-year-old infants that had been confiscated from Zairean smugglers and arrived together in a tiny box. Although Girly was starved and dehydrated, she was actually in pretty good health. But Junior was in a ghastly state. He was suffering from a very high temperature, severe dehydration, and an inconsistent pulse, and was so inert that he seemed to be in a coma. He also had bloody diarrhea and a gunshot pellet embedded in his right arm. It took weeks of hourly medication and bottle-feedings before Junior came around and began to gain strength, but even though he recovered sufficiently to join the other chimps, he never seemed quite right. The bullet wound and lingering internal injuries left me worried that he might never fully recover from the wounds he received during his capture.

Girly, meanwhile, was adopted almost immediately as a little sister by Liza. In fact, Liza carried Girly everywhere like a baby, either cradling her in her arms or slinging her up on her back and allowing her to ride like a jockey out on the daily bush walks, and Liza would make nests high up in the trees each day and curl up alongside Girly for naps. As gratified as Girly surely was to be given such attention, I wonder if Liza didn't enjoy it more. Her excellent maternal skills were a sure indication she'd come from a family with a strong mother and very likely many brothers and sisters, and gave me hope that one day she might be a successful mother in her own right.

Our first four chimps were all infants and roughly equals, but the next chimp to arrive changed our group dynamics dramatically. Charley was an adolescent male who came to us in June 1984, at the age of six. He'd been kept as a pet illegally in Zambia for about two years before his owners brought him out, and was in excellent health. But as sociable and playful as he was with the younger chimps—and though technically still a youngster himself—Charley was large and very powerfully built, and he quickly assumed the role of dominant male. He even claimed the oldest female, Liza, as his private property, in no uncertain terms. We'd never had a hierarchy before at Chimfunshi, yet there was no denying that Charley was the new boss and that the others did well to respect him.

The last of the six to arrive was Bella, a sweet-natured and healthy two-year-old female who came in August 1984. Like most of the others, she'd been confiscated by Zambian game rangers, but the fact that she was in such good condition made us think she'd probably been kept in a private home. Bella struck up a close friendship with Girly, and joined with Liza to form our first female "clique." Jane Goodall's book told how females establish close social bonds in the wild, regulating everything from family movements to male dominance, and it was pleasing and encouraging to see that our females could assume such roles so soon at Chimfunshi.

But now that we had six chimpanzees, the notion of allowing each to sleep in our bed or sit with us at the dinner table seemed laughable. Dave was forced to build large outdoor cages for the chimps alongside our house, with wooden boxes on tall ledges that they could climb into for sleeping. Of course, what really interested the chimps was lining up in a row and watching us through the windows of the house, especially at night, when the lights attracted all six sets of eyes like moths to a flame. We also had to hire extra help to take the chimps on their daily walks in the bush, and that process proved terribly hard. Given that Dave and I had entirely improvised up to that point, it was difficult to know what to look for in a chimp keeper. But we found several young men from the nearby villages who appeared eager, and who seemed to take to the job quickly enough.

But simply hiring more staff did little to address the real issue facing Dave and me: Whether to go forward as chimp keepers or to instead find these chimps a permanent home. We had no grand plan in those days, but we knew enough to think that maybe treating the chimps like chimps was the best way to go. We also knew enough to ask for help. So I sent a letter and some simple drawings I'd made of my chimps to the National Geographic Society in the United States and asked if they could help me. I figured they knew an awful lot of people, and I hoped they could put me in touch with somebody who could give me some advice. Thankfully, a lady at the National Geographic wrote back and supplied me with some addresses, and one of them was that of Dr. Shirley McGreal, the president of the International Primate Protection League (IPPL), in Summerville, South Carolina. I wrote Shirley a letter explaining that I wanted to do what was best for my chimps, and she took my appeal to an IPPL meeting and asked for help. Before long, letters began pouring in from IPPL members, telling us what to feed the chimps and how to treat their injuries and so on, all of which was extremely helpful.

The most intriguing letter we received at that time was from Eddie Brewer, an Englishman who along with his wife, Lil, and daughter, Stella Marsden, had founded a sanctuary for orphaned chimpanzees in the West African nation of Gambia. The Brewers' sanctuary had been up and running since the 1960s, and the particularly intriguing aspect of it was that Stella would take the chimps that had outgrown the nursery and reintroduce them back into the wild at the Niokolo Koba National Park in Senegal. She'd had particular success with chimps that had been born in the wild and kept only briefly in captivity, and Dave and I were thrilled at the thought that our chimps might get a second chance at living like wild chimps. So, after an exchange of letters, Dave and I went to Gambia in early 1985.

The Brewers' sanctuary was an amazing place, situated on an island in the middle of the River Gambia. The chimps could live naturally, climbing trees and playing in the forests, but would not attempt to escape since chimps cannot swim. (Although they love water, chimpanzees' bodies are too dense with bone and muscle to float, and they sink like stones whenever they get in over their heads.) Once a day, a boat went out to the island to set food out for the chimps, but they were otherwise left pretty much alone, to behave as they would in the wild. As wonderful as the sanctuary was, however, and as politely as Stella treated us, the sad truth was that she could not accept our six chimps. The island was already close to full, she said, and introducing six youngsters to the group would almost certainly cause fights and would likely end badly for our chimps.

Reluctantly, Dave and I returned home and faced the sobering truth: we were chimp keepers after all. Not that this was a bad bunch. When I look back on those days now, I can honestly say we began with a tremendously well-behaved group of chimps—which was fortunate, since Dave and I were learning as we went along. But as the Zambian game rangers kept up the pressure on illegal poaching and smuggling, our numbers continued to swell, and four more baby chimps arrived in 1985, forming the foundation of a nursery group we called The Youngsters.

First was Spencer, a young male who had been brought illegally into Zambia as an infant by his owners and who lived as a family pet for about a year. Spencer was about four years of age by the time he arrived, but he bore the telltale scars of a wild-caught chimp: One finger on his right hand was badly deformed and he had a terribly weak left wrist, probably as a result of an injury sustained during his capture. But Spencer was still one of the handsomest chimps I have ever seen, with a very strong, classical face. He quickly settled in and became extremely protective of the other young chimps in the group.

Tobar arrived next, armed with more of a personal history than most of our orphans. He was bought from a hunter in Liberia for $15 as an infant, was then kept in Monrovia as a family pet for about two years, and was approximately three and a half years old by the time he was flown to Chimfunshi. Tobar was the first chimp we'd ever received who we knew came from outside Central Africa, but even though he didn't look much different from the others, he was surely the most unchimplike. His every movement seemed geared to curry favor with humans, and he had great difficulty fitting in with the others. The others, in turn, seemed to go out of their way to ostracize him. Was it because he was from West Africa and somehow not like the rest of them? Or was it because he was so much more comfortable with humans? I couldn't tell, but whenever Tobar felt particularly insecure, he'd retreat to a corner and suck his finger, like a human child.

The next chimp we received was Boo Boo, another house pet who'd outgrown his house. Boo Boo was bought from a hunter in Zaire and smuggled into Zambia, but had only lived in a private home for six months, so he was still an infant of a year and a half when he arrived. He adapted quickly to the others at Chimfunshi, but also displayed an intelligence and a penchant for mischief far beyond the other chimps. He elected himself the leader of The Youngsters, and there were times his expression reminded me of the many statues of Buddha I have seen—so knowing and yet so inscrutable. Sometimes, when you called the others for food, Boo Boo would turn his back, fold his hands over his potbelly, and seemingly wrap himself in an air of mystery. He would remain like this until almost all of the food was gone, then slowly rouse himself and wander over to demand his share. And if all the food was gone, Boo Boo would first ask the others for some of theirs by politely cupping his hand under their mouth—and then, if none was offered, he'd get terribly angry. I have seen Boo Boo, arms flailing and up on two feet, forcibly grab food out of another chimp's mouth, and the resulting screams would be shrill enough to wake the dead.

The last newcomer was Donna, a one-year-old infant female who pitched up in wretched shape. She'd been confiscated by the Zambian police, who caught an animal dealer attempting to sell her for $300, and arrived suffering from malnutrition and severe dehydration. She was terrified of humans and would shriek and attempt to bite us whenever we tried to get near her; other times she'd spit at us or attempt to rip out clumps of our hair. Obviously, she must have been treated badly by her captors. Donna regained her health rather quickly, but her fear and anger toward humans took much longer to overcome, even though I'd spend hours on end sitting near her and quietly talking to her, letting her get used to me. She integrated well with the other chimps, however, and delighted in playing the "baby" out in the bush. In fact, there were times I wondered if she didn't fall out of low trees or whimper pleadingly for more food simply because she knew it would prompt the other chimps to rush up and give her attention.

Donna was particularly fond of Boo Boo, who seemed delighted to have a playmate so close to his own age, and their adventures often left me breathless from exhaustion and laughter. In 1985, we had a female bullmastiff named Georgie who had a litter of six puppies. They were beautiful and of course Georgie was very proud of them, but she was also very wary of the chimps' getting too near. As a result, the chimps wanted nothing more than to get ahold of Georgie's puppies. One day, Georgie left the puppies and went for a run in the forest, and it was all the opening Boo Boo and Donna required. They started teasing them and pulling their ears, and when I heard the puppies crying I rushed out and shouted at them, telling them how naughty they were and doing my best to make it clear that the puppies were not to be teased. They actually seemed to listen to me, and went off to play somewhere. But once they decided I wasn't looking, they doubled back.

Georgie must have heard her puppies crying as she returned from the forest, for the next thing I saw was a pair of chimpanzees running for dear life from a furious mother dog. She chased Donna and Boo Boo round and round the garden until eventually they ran up a tree. Unfortunately, in their haste to get away, the chimps didn't look where they were going and scampered out along a dead branch. Their combined weight was too much for the branch and it came down with a resounding crash, chimps and all. To make matters worse, the chimps fell right into the middle of a column of army ants, which wasted no time in attacking and biting them. Boo Boo and Donna began jumping up and down, trying to shake the ants off, while Georgie stood and barked, wagging her tail as if pleased with the general outcome. I spent a long time pulling ants off Donna and Boo Boo, but I think the escapade taught them a lesson. They never went near the puppies again.

At the time, many experts we corresponded with told us that we couldn't expect to successfully integrate chimps from different parts of Africa. They said a Zairean chimp would never get along with a Liberian chimp and so on, but despite the initial ostracism of Tobar (he finally became close to Spencer and several of the others) we did not find that to be true. I think our chimps recognized they were all from different families, and they all certainly looked different—some had lighter skin than others, some had freckles, some had more or less hair—but that never stopped them from living together happily. And Dave and I began to wonder if maybe the differences between chimps weren't more environmental than genetic. One study we read said that chimps in Liberia were distinguished from those in the Congo by the fact that they used rocks to crack open nuts, whereas chimps in the Congo didn't know how to crack nuts. But what it failed to mention was that there *were* no nuts in the part of the Congo that the study focused on. And I feel certain that if you'd given any of our chimps at Chimfunshi some nuts, it wouldn't have been long before they found *something* to crack them open with—regardless of where they came from.

Our chimps seemed to recognize each other as individuals and as a species, and that seemed to be enough for them. And if two or three or ten chimps that had had hard lives could come together and support each other at Chimfunshi, that was enough for me. Like humans, chimps need time to get to know one another, and we paid careful attention as each of the relationships grew, especially in the first few days. But there's no denying that the chimpanzee "community" we created owed its success equally to instinct and luck.

As our chimp family grew, so did the number of special friends who generously dedicated themselves to Chimfunshi, and it's no exaggeration to say we could not have managed without them. It was at the Brewers' sanctuary in Gambia that we first met a wonderful Swedish woman named Ingrid Regnell, who was visiting there and with whom we had long discussions about our chimps and chimps in general. She was absolutely passionate about animals. About a year after we met, Ingrid wrote and asked if she could come out to Chimfunshi for a visit. Of course we said yes, but her first trip proved disastrous. Ingrid fell ill one night and began suffering from terrible diarrhea, and became so sick that we rushed her to the hospital in

Chingola. There, the doctors discovered that she had contracted shigella, a form of typhoid that is spread through infected water, and we were all very worried, as we had no idea how long she'd been ill or how deeply she'd been infected. But with lots of treatment and care, Ingrid slowly recovered—even though the Swedish health authorities were forced to immediately escort her off the plane for testing when she returned home.

We never thought we'd see Ingrid again after that horrible experience, but she had a travel insurance policy that paid her enough money for another plane ticket to Zambia, and she courageously returned about a year later. It was during that second trip that Ingrid offered to help us raise money for Chimfunshi through an adoption program, whereby donors would "adopt" our chimps for a fee and receive regular updates on their progress. We do not know what made Ingrid so successful, but she single-handedly raised thousands and thousands of dollars for Chimfunshi—all at a time when the chimps were becoming a serious financial burden for us—and developed a network of supporters that exists to this day.

Not long after, we received a visit from the British author Jack Stoneley, who once wrote a very popular and heartfelt book called *Tuesday's Dog*. I had been publishing a small newsletter myself for children that included stories and games about chimps, and I thought it would be wonderful if I could serialize *Tuesday's Dog* in my magazine, so I wrote to Mr. Stoneley's publishers for permission. I said I couldn't afford to pay for the rights, but would certainly give him free copies of the magazine, if he wished. A few weeks later, I got a garbled message over the short-wave radio from my daughter Diana, who said, "Mr. Author is arriving on Friday from England and wants to come out to the farm." I had no idea who "Mr. Author" was until Friday came and Jack Stoneley stepped off the plane. I was shocked—I had no idea he'd received my letter, let alone planned to visit. But it seems he was writing a book about chimps and was seized by writer's block when my letter arrived.

"Betty! This is an omen!" he'd shouted to his wife. "I have to go to Africa at once!"

Jack was a wonderful man who had an excellent rapport with animals, and he got on very well with the chimps. He borrowed a video camera while at Chimfunshi and made a video that he later put to music. Some television people back in Great Britain apparently saw the video and were so inspired that they came out to film their own documentary for the popular program *Nature Watch,* and the resulting two-part film, *Chimps in Crisis,* told our story to the world. Suddenly, Chimfunshi was not just our little cattle ranch—it was the Chimfunshi Wildlife Orphanage, and inquiries and support began arriving from all over Europe and the United States.

The chimps, meanwhile, were demanding more and more of our time. The daily bush walks we took had been initiated to give Pal and his cohorts a bit of exercise, but also to make sure they remained accustomed to their natural habitat. We encouraged them to climb trees and tip over anthills and eat wild fruit, simply because that is what their mothers would have done with them in the wild. I sometimes heard people refer to the bush walks as our "teaching" The Youngsters how to behave like chimps, but that's rubbish. The chimps always knew exactly what they were supposed to do in the bush; all we ever did was give them the opportunity to practice. If anything, they taught us more about life in the bush than we did them, and if ever I was stranded or lost in the forest with my chimps, I'd simply watch them and do what they did to survive.

The bush walk routine developed over time, until we settled on a schedule that felt about right. Each morning at around seven o'clock, after The Youngsters had their cups of milk, one of the keepers would load up a bag with fruits and vegetables—things like figs and green peppers and bush oranges that chimps really enjoy—collect the chimps from their cages, then set off into the forest northwest of the farm. Each of the chimps insisted upon being carried, of course, so quite often it took two or three of us to get them all into the bush, but once out in the wild, the chimps scampered down and walked on their own. For the next seven hours, the chimps would be led around the forest on our property, stopping every so often to snack on the goodies in the bag, but mostly roaming and playing as they pleased.

Around midday, the chimps usually settled down for naps, and most began making nests in the trees. Some, like Boo Boo, were quite good at this, but others were not so good and were forced to learn by first watching, then slowly duplicating the complex weave of limbs and leaves and branches, all held down by the chimp's own weight. There is quite an art to making these nests. The chimps first seek out a strong, firm place in the tree, such as two horizontal branches or the crotch or upright fork that forms the foundation. Then they reach out for smaller branches, which are bent onto this foundation, some from the right, some from the left, some from in front. Each is placed on top of the other and held there with a foot while the next branch is selected. Then they tuck in the leaves at the side, often breaking off a few leafy parts to make a pillow for their heads. Most chimps, after lying down for a few minutes, sit up and rearrange some of the leaves, perhaps attempting to smooth over a rough spot here or there; but I have sat on one of these nests and am amazed at how strong they are. Once settled, the chimps will nap for anywhere from forty-five minutes to two hours.

The bush walks were intended to expose the chimps to as much nature as possible, and they thrived in the forest. We'd normally walk slowly, with the chimps winding casually in and out of line behind us, but when one of them spied a fruit tree off in the distance, he'd grunt to the others and they'd break ranks, scampering off willy-nilly for the tree and sometimes spending hours gorging themselves on delicacies like figs or cherries. We might encounter animals such as bushbuck or antelope in the forest, and now and again a herd of our own cattle. It was unusual to see snakes, though, because they usually slithered out of the way when they felt the vibrations of our steps as we approached. But when we did meet up, the chimps' reaction was always a strange combination of curiosity, fear, and a tremendous amount of caution. I have seen a chimp move within ten feet of a little snake, blow up its fur as big as it could, puff out its chest, and generally try to seem as scary as possible, barking and shouting at the top of its lungs. But the snake or the chimp would usually retreat before they got too close—there is much to be said for instinct.

As more chimps arrived and my responsibilities on the farm continued to take a toll on my time, I was not always able to go out on the bush walks with the chimps. But when I did, I felt privileged to be so close to their enjoyment, their antics, their tantrums, and their tears. It takes tremendous patience and energy to care for a half-dozen baby chimps properly, and we were very lucky early on to have found a young man named Patrick Chambatu, whom we hired as one of our first chimp keepers. Patrick, who came from one of the local villages near Chingola, was wonderful with the chimps and really seemed to enjoy his time with them. He also had what I call "chimp sense"—he understood how to read the chimps and how to anticipate problems, and he came to appreciate their personalities and special needs. For almost seven years, Patrick was the primary baby-sitter for the chimps, working his way through several waves of new babies and orphans, and became increasingly respected among the primate community for his work.

One of the most valuable things Patrick did was keep a daily diary on his bush walks, and he used to delight in reporting back to me each day on the latest developments and escapades of the chimps. For instance, on cold mornings, Patrick would stop in the forest and make a small fire to warm his hands. It wasn't long before Tobar—who probably saw fires being made by his owners in Liberia before he came to us—began to take it upon himself to go off and collect firewood for Patrick. He would bring back an armful of sticks, drop them near Patrick's fire, and go off for some more. Then he would imitate what Patrick was doing and hold his hands out to the fire to warm them. But this practice came to a halt the day Patrick saw Tobar sneak up to the fire, look around very carefully to make sure no one was watching, pull a small stick out of the flames, and creep away with it. There was a little pile of dried grass behind a bush, and Tobar stuck the glowing end of the stick into the grass, obviously trying to start his own fire. Fortunately, the grass did not ignite, and Patrick was able to wrestle the stick away. But after talking this over, we agreed that Patrick would not light any more fires in the forest.

After being out all day, The Youngsters would come back home to a meal of vegetables and fruit, which included oranges, pineapples, bananas, watermelon, guavas, mangoes, avocado pears, and anything else that was in season. They were also particularly fond of sweet potatoes, leeks, lettuce, cabbage, pumpkins, tomatoes, raw garlic, onions, and green beans. One of the chimps' favorite foods was peanuts; they'd spend hours casually sitting in a circle and cracking the shells, grunting contentedly to themselves, like old men sitting around and telling each other stories. Each of the chimps had personal likes and dislikes, but there is such a variety of fruits and vegetables in Zambia that we usually found something to everybody's liking. After eating, they played in their cages until about five o'clock, when it was time for their cooked maize, a thick porridge made from cornmeal. We prepared this with water, salt, and vegetable oil, slowly boiling it in a large cauldron over a low flame until it congealed into a paste. Once the mixture cooled, we'd roll it into balls and give each chimp three or four, which they'd take to their favorite corner and eat slowly. By then it was bedtime, and the chimps would climb up into the old tires we'd placed on the shelves in their cages—always two per tire—and cover themselves with burlap sacks and fresh straw before falling asleep.

During this time, Liza Do Little recovered so quickly from her initial injuries and trauma that it defied description. In a little over a year she was bigger and stronger than any other chimp except Charley, and it became increasingly difficult to keep her in cages with the other chimps. Liza was a specialist at making small holes in the wire mesh, then squeezing herself through, and would always head straight for the nearest trees to build big nests to nap in during warm afternoons. Liza and

Charley's relationship also grew more complex as they approached maturity and sought to distance themselves from The Youngsters. A chimpanzee starts to reach adolescence at around the age of eight, but does not really become socially mature until about twelve. A young chimp is lovable and manageable and craves the affection of its mother, who breast-feeds until her baby is about four years old. Late in the fourth year, the mother often becomes pregnant again and the youngster graduates into a social group of other chimps approximately its own age. Once a chimp passes the age of four, its physique begins to change dramatically, as muscles fill out the shoulders, backs, and thighs, and its independence emerges. In captivity, that means the keeper's word is no longer law, and a desire to test the limits of authority becomes a dangerous game.

At the time Charley reached adolescence, he had been nothing but a model chimp with a lovely character. Despite his massive shoulders and obvious strength, he was never aggressive and seemed to genuinely like Patrick, who had been accompanying him into the forest six days a week for over three years. But one day, as Patrick was returning home from a bush walk with four of the older chimps—Tobar, Spencer, Charley, and Liza—Charley suddenly veered about a half-mile from the house and noisily chased Liza off into the trees. Patrick approached Liza to try and coax her to continue home, but Charley had other ideas. He stood up on two feet, blew himself up to impressive size—his every hair standing on end—then rushed at Patrick and knocked him down. Poor Patrick was badly shocked, and although he tried, he could not get either Charley or Liza to follow him. He returned home with Tobar and Spencer, then immediately went back out in search of the two others, but he could not find them.

The next day, Patrick went out early and found signs in the forest indicating that the chimps had tried to come home, but had apparently been frightened off by a bad thunderstorm. We were growing quite concerned, as none of our chimps had ever behaved this way before, but late that afternoon Patrick finally found them in a grove of trees about two miles from the house, and they followed him home willingly. Liza actually appeared pleased to be back, and Charley ate like a horse. But the episode left Patrick badly shaken, and all of us were on edge, wondering when Charley might fly off the handle again.

Meanwhile, the chimps just kept coming. Chiquito was next, a four-year-old chimpanzee that had gained a measure of local fame as a tourist attraction at a Zambian resort near the town of Mufulira. In fact, Chiquito had been the first chimp I ever saw. I had driven out to Mufulira in mid-1983 to have a look at this chimp everybody was talking about—before Pal ever pitched up, of course—and Chiquito made a very deep impression on me. For one thing, he looked to be about one year old at the time, but his body was almost entirely devoid of hair. His owners also dressed him in baby clothes, which gave him an odd appearance of something half-human, half-ape.

But as Chiquito grew, he got bigger, stronger, bolder, and more destructive, and an elderly woman who lived at the resort said it was like being a prisoner in your own home: She spent all day locking doors and windows to keep him from getting into places where he did not belong and destroying things. For these reasons, the owners of the resort gave Chiquito to a man who lived in Ndola, and he placed a thick collar around Chiquito's neck and kept him chained to a post in his yard. He also taught the chimp to drink alcohol and smoke cigarettes, neither of which made Chiquito any easier to control.

By the time Chiquito's owner contacted us and offered to send him out to Chimfunshi, Chiquito was a large and very powerful chimp indeed, weighing close to seventy pounds, with round, muscular shoulders that sat atop his blocky frame. He looked nothing like the little toy I saw dressed in baby clothes so many years before, and we were a bit doubtful that we could handle his transfer safely. My daughter Diana drove out to Ndola to collect Chiquito in her truck, and he rode back in the passenger seat alongside her all the way back—a trip of about three hours. We met them in Chingola, where the real trick was to get him out of Diana's truck and into ours. It was quite a struggle to convince the chimp to switch vehicles, even with all of us pushing and pulling on him, but it says a lot for Chiquito that we were able to manhandle him in that way and not get bitten, and he eventually consented to ride alongside me back to Chimfunshi—although I must admit I kept one eye on the road and the other on him the whole way.

Back at the farm, Chiquito let us know just how strong he really was. His collar broke as he was being transferred to his new cage, so we fitted him with one of our leather bullmastiff collars until we could get everything sorted out. But without using his hands, Chiquito calmly braced himself against a pole and flexed his powerful neck muscles, placing such incredible strain on the collar that it simply snapped and fell away. Dave and I were speechless. Luckily, Chiquito was fairly accepting of us, and he turned out to have a lovely nature, but the mere thought of that collar ripping like paper was enough to always keep me on guard.

Chiquito did present us with several distinct problems, however. First, he was addicted to alcohol and cigarettes, and we were forced to wean him off these vices the only way we knew—

cold turkey. It got so bad that we had to be careful not to go near Chiquito with beer on our breath or he'd smell it and become angry, and we asked visitors not to smoke within sight of him. But worse than that was the fact that Chiquito was petrified of the other chimps. Having been raised in human households, he did not know what to make of his own kind, and we were forced to keep him in a separate cage until he eventually got used to the way chimps sounded and smelled and looked.

One month later we received Tara, a young male chimp about three years of age who also enjoyed local celebrity status. Confiscated by the Zambian police from smugglers in Ndola, he was found in a tiny cage, so cold and neglected that one game ranger said he was "like a block of ice." Tara was taken to a private home in Ndola and wrapped in blankets, then tucked into a bed with hot-water bottles. It took nearly twelve hours for his body temperature to come up close to normal, during which time he refused to eat or drink. But he eventually began taking sips of milk, then ate a bit of banana, and finally began taking an interest in what was going on around him. Still, Tara was not able to join us at Chimfunshi until he appeared in court at the trial of the smugglers as Exhibit A.

At the Ndola magistrate court, Tara put on quite a show. He attempted to shake hands with one of the public prosecutors—not once, but twice—and was so engaging that even suspects and accused persons in nearby cells were roaring with laughter. The *Zambia Times* included a write-up on the case under the headline, "Chimp Plays Tricks," along with a quote from the public prosecutor: "That thing is like a human being. I think in the past human beings were like that. But I feared (*sic*) to shake hands with it because it was the first time I saw an animal with human features."

Tara's antics were largely in vain. The smugglers got off with a very light sentence, and we were left with a three-year-old malnourished chimp that weighed just eight pounds upon arrival. Tara was too weak to be placed in with the other young chimps, so we let him sleep in bed with Dave and me, where he did something I'd never seen a chimp do, and haven't again since. As we had stopped potty-training chimps after Pal and insisted on making those who slept with us wear diapers, Tara took it upon himself to alter the equation. When he had to use the bathroom, he would wriggle out from under the covers, walk to the foot of the bed, pull down his diaper, and stick his rump out over the edge of the bed to go. When he was finished, Tara would pull his diaper back up and crawl back in between Dave and me.

We took turns watching and caring for Tara during the day, and he quickly grew, from strength to strength, becoming fit and strong enough to join the others after just three months. Tara established himself as a complete clown and an extrovert, spending most of his time galloping around while turning funny, half-backwards cartwheels, and whenever we had a visitor, Tara was always the first to rush up and offer to shake hands.

Four more chimps arrived over the next seven months—Cleo, Rita, Cora, and Sandy—bringing our chimps to a grand total of sixteen by the end of 1986. But it was clear to Dave and me that our system of cages and bush walks was not designed to accommodate so many chimps. Nor were we comfortable simply building cages and stuffing chimps inside them—when you do that, you've got a zoo, and that was never our intention at Chimfunshi. We were also beginning to receive "misfits" now and again, chimps such as Chiquito who did not easily fit in with either the older established group or the newest wave of youngsters, but who nevertheless required special care and attention. And who knew when Charley would finally blow up and behave in the aggressive, demanding, violent manner that an adult male chimp should? All of these concerns began to weigh heavily on us, and Dave and I knew we would have to make some changes.

Five

The Great Wall

The chimps look out over The Wall from the treetops.

It was on the plane back from visiting the Brewers' sanctuary in Gambia that Dave first began making sketches. We had both assumed that Gambia would be the logical destination for Pal and Liza and the rest, and when it became clear that the Brewers couldn't accept our chimps, the sense of sadness and disappointment we both felt was very real. We were also terrified—we hadn't really imagined keeping these chimps forever, and now that looked like exactly what had to happen.

Fortunately, Dave's background as an architect and a building contractor saved the day. He began drawing enclosures and cages and all sorts of possibilities, just rough outlines really, but his brain was clearly working on something big. He'd make a quick sketch and then show it to me, and then we'd find flaws in it and he'd try something else. It went on like this for hours. It was obvious that the Brewers' concept of an island—basically a cage without walls—was a wonderful solution, and we agreed that it was exactly what we'd hope to duplicate at Chimfunshi. But there were no existing islands in our section of the Kafue River, and the river itself was not wide or deep enough to sustain a man-made island, so Dave said,

"Why don't we build an island on dry land?" He came up with the idea of digging a moat around one of our pastures, thereby giving the chimps plenty of space while ensuring that they would not escape.

Unfortunately, it took just one survey of our property to make it clear that a moat was out of the question. Chimfunshi may seem fairly flat and stable, but the land we wanted to use turned out to be seventy feet higher in some places than in others, and the proximity to the river left the ground soft and unreliable. So that was no good, especially since our minds boggled once we did the math and realized the expense of building a moat. We began chasing ideas, and every day our predicament got worse. We already had sixteen chimps, but by the time we actually got something—*anything*—built, who knew how many more would've arrived? Dave, however, was undaunted. If we couldn't dig down into the earth to create an island, he reasoned, why not build a wall instead. Thus was created the Great Wall of Zambia.

Dave's plan was simple: If we used the southern bank of the Kafue River as one border and then erected a wall around the other three sides, we could enclose a large area with access to trees and open grassland and even the river itself, since we already knew from the Brewers' experience that chimps would not be able to swim beyond the wall. Dave did some figuring and settled on a seven-acre plot of pastureland just a few hundred yards up the main road from the house, with some fruit trees and thick undergrowth. The riverfront would form a natural boundary of two hundred yards, with the wall encompassing another three sections of two hundred yards each, making a rough square. Dave then set about configuring what sort of wall might best keep chimps in. Electrical fencing was not an option at that time, but Dave visited a few zoos in Zambia and elsewhere and looked at a lot of books and drawings, and he eventually came up with a design of smooth brick and mortar that was ten feet high, since we were told that chimps couldn't jump higher than eight feet. At the front, an interconnected maze of concrete handling cages with steel bars and steel sliding doors would be built in case we needed to bring chimps inside for feeding or sleeping, and a large sliding steel door would be built into the wall to give easier access to the enclosure itself.

Once word got out that we were going to build a walled enclosure for our chimps, everybody thought we were crazy. Especially the ape community—oh, they were *sure* we were bloody crazy. And seven acres? Nobody had ever built anything that big for chimps before. Lots of people dropped by to check on our progress, and I'm afraid most of them left shaking their heads. Maybe we *were* crazy, but we really couldn't think of anything better.

Luckily, we found some wonderful friends who supported our efforts and kept our doubts at bay. I got a letter one day from an American woman named Margaret Cook, who asked if she could come to visit and get a "chimp fix." Margaret had worked at the Kansas City Zoo and had lots of experience hand-rearing animals such as chimps and gorillas, and had heard about our sanctuary and was anxious to be of help. Her husband, Thomas, was a retired doctor, and I invited them both to come. They were in their sixties and really a wonderful, fun couple, and they did so much for me and my chimps. Margaret brought all these zoo charts for feeding and medicating baby animals, and they taught me how to recognize and treat certain illnesses. Thomas also gave me my first stethoscope, which I continue to use.

The Cooks' visit coincided with that of a young German man, Stephan Louis, who managed a successful motorcycle company in Hamburg but was passionately devoted to animals. Stephan has since become a Chimfunshi trustee and one of our most dedicated fund-raisers, but we really didn't know what to expect when he first arrived, or if he and the Cooks would even get along. All doubts vanished early one morning, however, as Margaret and I were sitting in the kitchen and the door burst open. Charley had somehow escaped and he dashed into the room, knocking Margaret's cup of coffee out of her hand and scaring us both to death before galloping off toward the bedrooms. I ordered Margaret to go and lock herself in the nearest room—which happened to be Stephan's—and she ran off screaming, "Stephan! Stephan! Open the door! Charley is out!" But Stephan had been in the shower, and as he raced to the door stark naked, he met up with Margaret. Margaret came to a dead stop, composed herself, and said, "You silly boy! Get some clothes on! Charley is out!"

Another good friend who joined us around that time was Carole Noon, a bright young American primatologist whom we eventually came to regard as an "adopted" daughter. Carole's love and understanding of chimpanzees is astounding, and she returned again and again to Chimfunshi, sometimes to conduct behavioral studies on our chimps on topics such as habitat enrichment and resocialization, sometimes just to be among friends. Carole now has her own chimpanzee sanctuary for "retired" NASA astronaut chimps in Florida and has become one of the most respected experts in the field, but I like to remember the times we'd take concrete blocks up onto the roof of an old shack here and sit and drink beer together as the sun descended over the floodplains.

Work on "The Wall" finally got under way on March 10, 1987, and right off we could see this was going to be slow-going. For starters, each brick had to be made by hand, since Dave wanted the wall to be twelve inches thick and no bricks came in that width. It took one bricklayer, a wonderful Zambian named William Chinyama, two years to make and lay all the blocks, and work was held up for long periods during the rainy seasons, from October through April, when the mortar would not dry or the ground was too soft and wet to steady the foundation.

Inflation was also a terrible problem. When we started in 1987, a bag of cement cost about 80 kwacha, which was very cheap, but the price had risen to almost 30,000 kwacha in a little over two years' time. Zambia's economy virtually collapsed in the late 1980s, when the copper market went bust and inflation sent the price of ordinary items sky-high. The government countered this by issuing new money, but the problem was, you were only allowed to exchange 2,000 kwacha at one time—which was a pittance—so everybody was forced to resort to a barter system and panic was rampant. It was a very anxious time because so many of our friends pulled up stakes and left Zambia then, and we kept thinking the economy couldn't get any worse, but then it would. If it hadn't been for duty-free shops and the generosity of our overseas supporters, Dave and I might have been forced to give up, too. Dave used to joke that he was only staying as long as the beer held out, but ordinary items such as bread, sugar, matches, and cooking oil were extremely scarce, and a cheese sandwich came to be regarded as a tremendous luxury. I remember going into the Bata shoe store in Chingola one day and the place being absolutely empty, except for four staff members and a couple of mismatched pairs of plastic children's shoes.

In fact, things got so bad that the police began arresting women who queued up too early for goods at the government-run shops. The lines began forming at three o'clock in the morning—even though the shops did not open until late afternoon—but the police were afraid that thugs would molest women at that hour of the night, so they arrested the women for their own safety. Meanwhile, a group of Australians visiting Chimfunshi then presented Dave and me with a gift I doubt either of us will ever forget: six pounds of kippers. Just the smell of them frying at breakfast made us both laugh out loud, and we ate them slowly to make them last. They were delicious.

Even during the worst of times, however, work continued on The Wall. Though the cost of cement and sand was spiraling out of sight, having committed to a mortar wall we had no choice but to pay the going rate and forge ahead. On several occasions, we were forced to sell some cattle to meet the bills—a step that is never taken lightly by cattle ranchers—but we also tried everything we could think of to raise money for the project. A music teacher in nearby Kitwe who had visited the farm organized a concert by the Lechwe School orchestra, and it sold out the Kitwe theater, raising 5,000 kwacha in the process. We also held a dance in Kitwe, and I seem to recall more than a few bake sales, and many people were very generous.

Finding enough steel for the bars and doors in the handling cages was difficult too, as we had a steel shortage in Zambia at the time, but Carole Noon and I managed to solve that problem on our own. A few of the mines were all still active, and there were scrap yards filled with broken bits of machinery and vehicles that the mining companies couldn't use anymore. So Carole and I drove around with my little truck and picked up odd chunks of steel from scrap yards, then took them to get cut down into the pieces we needed. Or we found old drill bits that could be welded into the bars for the cages. If we'd had to buy that much steel—or even pay the going rate for it—it would have been impossible.

I remember there was one particular scrap yard in Chingola where I spoke to the owner and asked if we could look at his scraps and see if anything was useful. He seemed very busy and rather preoccupied, however, and I wasn't sure he even understood what I was asking. But when I inquired as to how much he wanted to charge for it per pound, he looked up quickly and said, "No, no, no, you can have what you want. Just take it away."

This scrap yard turned out to be enormous. It was probably as big as our whole compound at Chimfunshi, and Carole and I could barely contain our excitement. He led us into the main yard, and you'd have thought we were at the supermarket, the way we acted. Carole and I were saying, "Oh, that's nice" or "Isn't this piece lovely?" or "Have you ever seen such nice scrap?"

Finally, the owner said to me, "Look, you can go through all that if you want, but you'll have to excuse me because I've got an appointment. Take what you'd like. And promise to lock the gate when you leave."

He turned to go, and Carole leaned over to me and whispered, "That man is a fool. *I* wouldn't leave us alone in here."

We took all we could lift, but some of it was so heavy we couldn't lift it onto the truck, so we actually came back the next day with a seven-ton truck. And would you believe the silly man did the same thing again?

"I've got to run," he said, "but when you go, please put the padlock on the gate." And he was off.

We got a lot of scrap off him indeed.

There was never quite enough steel or cement or even manpower to keep The Wall on schedule, however, and by September 1988 we were running far behind. In fact, things got so bad that Dave began to constantly redraw his plans, changing the design of the wall and the cages to accommodate the raw materials he had at his disposal. And as our chimps grew larger and stronger, Dave modified The Wall by adding a strand of solar-powered electrical wiring along the top just in case they *could* jump higher than eight feet. But when The Wall still wasn't finished in March 1989, Dave and I resolved not to speak of it to people, since all they ever asked was, "When will The Wall be finished?" and we got sick and tired of saying, "Soon, soon." We began to wonder if that bloody thing would ever get done, and just as we predicted, the chimps kept arriving, with no letup in sight.

Little Jane came first, in July 1987—yet another chimp who had been confiscated by game rangers after being found tethered outside a local bar. We think she was about two years old when she arrived, but she weighed just ten pounds and was much smaller than the other chimps. She also had a very dark face, almost black, like a gorilla's. Jane clearly feared humans—I assume she'd been mistreated in captivity—but she took to the other chimps straightaway. It was as though she'd known them forever, and we were able to integrate her with The Youngsters in just a couple of days. She also exhibited better "bush sense" than many of the others on the daily walks, and seemed to know exactly what barks and resins and plants to eat. In fact, the others began to try many of these foods only after watching Little Jane eat them. Her tree nests were also the best of the lot, a clear sign that she'd been taken fairly recently from the wild.

But Little Jane's most amazing skill was her ability to climb trees. When out in the forest, she would spend most of her time in the highest branches of the trees, looking this way and that, and I sometimes wondered if she was just gazing out or if perhaps she was looking for something she recognized. Maybe she thought she'd find her way home. She'd first look very hard off in one direction, then turn around and look toward the horizon in another. Little Jane would do this for long periods each day, then eventually give up and settle into the uppermost branches alone and sit quietly as the tree swayed in the breeze. But I was amazed the first time I saw her jump from one tree to another. Chimps are normally cautious climbers when not in a hurry, taking time to select a strong branch before leaving the one they are on. But not Little Jane—when she leaped from one tree to another, she appeared to be flying through the air, and at first I thought she would not make it. But she'd land in the foliage of the next tree with a tremendous crash, then immediately start eating the leaves as if nothing at all had happened. It would occur so fast that at first I thought I had imagined it. But she repeated the trick on numerous occasions, and eventually we came to accept that as Little Jane's preferred mode of travel.

The next chimp to arrive returned from the dead. Coco was a young female who came to us in a terrible state—a semi-corpse, really. She had belonged to a smuggler in Kitwe, who brought her into the local SPCA when he thought she had become sick. The attending veterinarian treated Coco for a skin disease, but told the man to bring her back in three days for surgery, thereby giving himself time to contact the authorities and have the owner arrested. But the smuggler got confused, and returned three *weeks* later with poor Coco, who by this time was so ill that she was barely breathing. The owner laid her limp body on the table and said, "Please do something for her." Then he left. The vet examined the chimp and, as he could find no heartbeat, declared her dead. But a nurse who was holding Coco's hand suddenly shouted out, "I can feel something like a pulse!" Even though they managed to resuscitate the chimp, she was still extremely ill when she reached Chimfunshi, and required a month of intensive nursing just to be strong enough to sit up and walk. Coco recovered quickly after that, and it wasn't long before she was playing with the other chimps and going for rides on their backs.

Jimmy came next, a male chimpanzee who had been raised as a pet by a family in Kenya and was flown to Chimfunshi by his owners. Although happy and secure with humans, Jimmy proved to be another chimp who simply did not know what to make of his own kind. He was quite scared of the others, and his lack of bush sense and chimp savvy made him quite the odd duckling. At mealtimes, for instance, Jimmy was so used to being fed his daily portion that he did not understand how to compete for food. So rather than grab what he could and hoard it from the others—as most chimps do—Jimmy would carefully select one piece of fruit at a time, eat it thoughtfully, then look around and see what else there was to eat. Of course, the food would be long gone by that point, so we took to keeping some food aside for Jimmy and slipping it to him after the others had eaten their fill.

The next two chimps were a pair of females who came together, Big Jane and Josephine, who had been kept as house pets by a farmer in Zambia. Eight-year-old Big Jane was terrified of humans and reacted to any display of kindness with anger and aggression. But it was also clear she was not well. She had a lump about the size of a tennis ball protruding from her abdomen, and two smaller lumps on her lower back, down near her hips. We had her examined by veterinarians from the University Teaching Hospital in Lusaka, and she was found to have encapsulated abscesses on both hips and an encapsulated umbilical hernia. She was treated for her condition and improved considerably, but still walked with a slight limp. The veterinarian believed that Big Jane must have suffered the injuries either by getting kicked or being attacked with a weapon, which went a long way toward explaining her fear of humans. Six-year-old Josephine, meanwhile, quickly integrated with the other chimps, but displayed almost no bush sense whatsoever. It took many months for her to learn how to climb trees or forage for food, even though those were skills most chimps acquire by the age of two.

We now had twenty chimpanzees and, still, nowhere to put them. And other animals kept coming too; we experienced a sharp rise in protected species like African gray parrots and tortoises that had been confiscated by game rangers. We'd periodically release into the forests northwest of the farm "family" groups of a dozen or more monkeys that had been brought to us, but many of them preferred to stay close to the farm, or even refused to leave it at all. Still more animals were brought in by locals who'd raided the nests of owls and eagles and hornbills, hoping we'd pay them for the babies they brought us. We accepted the animals all right, but then threatened to inform the police. Most of the people vanished and never returned.

Meanwhile, a half-dozen completion dates on The Wall came and went, and as the delays stretched on into February and March and April and May of 1989, our frustration and disappointment grew to outright despair. Our patience was tested virtually every single day. If sand for mortar could not be found, then the sliding doors were not ready. Or the steel bars were the wrong size. Or the viewing platform needed reinforcing. And if you had all the raw materials, then a Zambian holiday was suddenly declared and the workers were nowhere to be found.

But little by little and day by day, The Wall took shape, stretching around the pasture and slicing through the forest just beyond the front gate. And then, finally, on Monday, June 19, 1989, the last brick was set into place. We were often told, "It cannot be done," but the so-called experts were wrong. We had really done it. The Great Wall of Zambia was complete, and we excitedly began making plans to place the chimps inside. We wanted to have an official party, of course, since so many people had done so much to help us on the project, but we were nervous about releasing the chimps into a new environment in front of several dozen strangers. Who knew what might happen? Who knew what might go wrong?

We decided to shift the chimps over to the new enclosure, bit by bit, giving them time to get adjusted to their new surroundings before the "official" opening, and we started right in with The Youngsters the following day. After their morning milk, we gathered up the chimps around 7:30 as usual, but instead of heading off into the bush for their walk, we headed for The Wall. Each of the chimps went in through the large steel door with surprising ease. We all headed to the river together and ended up at the large fig tree down by the bank.

Big Jane was the first one up the tree, but it was not long before the others followed her example. The Youngsters probably figured this was just another bush walk, except in a new direction.

But it was truly a delight for us to watch them romp and explore in this new world, knowing it would be their permanent home. Josephine and Jimmy played throughout the day and stayed close together, while Big Jane played chase with Rita and continued in her role as overseer of the group. She investigated the fights that broke out periodically among the other chimps and positioned herself on the door between the cages to keep an eye on both the indoor and outdoor worlds. Little Jane appeared much more playful with the others than she had been in the smaller cages next to the house, and Rita distinguished herself as the first chimp to pound on the large steel door in the wall with her feet and seemed genuinely pleased with the deafening *boom, boom, boom* it made.

But after the babies' evening meal, Rita suddenly became inconsolable. She whimpered and kept reaching out through the bars, and didn't settle down until after we had left for the evening. Rita's reaction to her first night in the new cage—out of sight of the house—seemed an appropriate response. This was a process of letting go, both for Dave and me and for the chimps. At the house, they were part of everything that went on, and from their cages they could see and hear so much of our lives. But it was time for them to grow up and move on, if only a short walk up the road, and the transition was hard on all of us.

As the official opening was still forty-eight hours away, I went into Chingola the next day to collect the mail and do a bit of last-minute shopping. I got back around noon to find that Dave had already moved Tobar and Cleo to the new enclosure, but we did not move any more chimps over right away. I stayed up very late that night to see where the chimps would sleep—in the cages or out in the trees—since this territory was so new to them. Surprisingly, Jimmy started making a nest right inside the door of the first cage on the left. He was awfully tired and he lay down to sleep, with Tobar hovering over him. Earlier on, Jimmy had tried to make Tobar go with him to the trees by pulling on his arm, and one time he took Tobar's hand and tried to pry it off the wire. But Tobar didn't want to go just then, and Jimmy eventually gave up. That night, Tobar lay down alongside Jimmy, and that is probably how they slept. There was only one other chimp in the cages, and that was Cleo. Then, just as it was getting dark, Rita sneaked into the cages and made a big nest in one of the tires. Out in the trees, I could make out a few shapes here and there, and it was lovely seeing Tara settle down into a big nest, but it was difficult to tell who was who in the fading light. But of that first batch, only four of the thirteen chimps slept in the cages, surely a positive sign.

Our optimism and delight at how easily the chimps accepted the transfer to The Wall was dashed the following day, however. We started the morning off very early by opening Pal's cage and simply inviting him to follow us to the new enclosure. He loped along happily at first, but then heard voices at the new cages and got confused. He started to turn back, but when he saw that Patrick and I kept going, he quickly caught up. Pal greeted the other chimps at The Wall through the wire of the cages and displayed quite a bit, puffing out his hair and making himself as big and scary-looking as possible as a means of establishing dominance, but then walked through the sliding door and into the front cage. Patrick and I then went over to fetch Spencer, and he followed easily and went in through the sliding door. We then went back for Bella and Girly, who raced over so quickly that they almost beat us to the new cages, and they also went in with no trouble. Next it was Chiquito—who took a shortcut, forcing Patrick and me to run up the road after him. But Chiquito would not go in through any of the doors, preferring instead to climb the wall from the outside and drop into the enclosure with the help of the electrical wire, which, thankfully, was not yet on.

We thought everything had gone extremely well and decided to call it a day. We did not want to move Charley and Liza until we were sure the electrical wire worked, maybe even wait until after the official opening, just to be safe. We were all standing around—Dave, myself, Carole Noon, Ingrid Regnell, and several others—congratulating ourselves and admiring the chimps from the viewing platform, when a terrible commotion started up from the house. Patrick came running over and said, very calmly, "Charley wants to come out." It seemed that Charley was trying to smash his way out of his cage and that Liza was also going mad. Clearly, they had seen the other chimps disappear and had no desire be left behind, so Patrick and I went over and opened Charley's door. Charley and Liza were out like a shot and also took a shortcut toward The Wall, with Patrick and me in hot pursuit. Liza disappeared somewhere into the trees, but as Charley approached the cages, he began slap-stomping the ground with his feet and puffed himself up to an enormous size. He raced into the handling room and displayed himself in the hallway between the cages, then slapped the wire on The Youngsters' cage and screamed at them. I had honestly never seen Charley so worked up, and, fearing that he might turn violent, I ushered everybody into an empty cage and we pulled the steel door shut behind us.

But Dave was still outside, and before I could warn him, he and Charley came face-to-face on the concrete path just outside the cages. Dave knew better than to turn and run, but he also had no desire for a showdown—so he froze. Charley swayed back and forth menacingly, his lips pulled back to display his teeth, then began walking toward Dave with exaggerated steps, as little swirls of dust appeared when his feet whapped the ground. It was clear this chimp meant business. Dave was in no position to defend himself, so, as Charley swaggered closer, Dave crumpled down to a sitting position and averted his eyes, the surest sign of submission in the ape world. And when Charley reached the spot where Dave was sitting, he paused for a moment, collected himself, sat down, and wrapped his massive arms around Dave in a bear hug. Then he kissed Dave on the shoulder, rose, and walked happily into the enclosure.

But then Charley suddenly went berserk. He began displaying again, and became so agitated and overwhelmed that he turned and threw himself at the first chimp he saw: poor little Jimmy. Jimmy seemed to sense the attack and tried to run toward the cages, but Charley caught him and threw him into the air, then jumped on him and dragged him along the ground. Most of the other chimps were screaming in terror, but Rita rushed at Charley and a couple of the others took courage from her and also came to Jimmy's aid, threatening Charley with hoarse "*wha*" barks at the top of their voices. The chaos and the noise were incredible. Charley finally dropped Jimmy and ran off at a slow trot, with, of all things, Jimmy running after Charley and threatening *him*. After a few yards, however, Jimmy wisely stopped and headed back toward the cages. He climbed up the front mesh and begged frantically to be let in, but before we could get to him, Charley reappeared at a dead run and yanked Jimmy off the wire, then threw him to the ground and stomped on him. The others rushed Charley as one, and he trotted off into the underbrush. Jimmy staggered to his feet and again begged to be let in. We managed to open the cage door this time, and Jimmy fell into my arms—just as Charley returned for another go at him. Poor Jimmy was shaking like a leaf and seemed to be in shock. His foot was pouring blood from a nasty cut, and he clung to me so tightly that all I could do was hug him back for the next half hour before he finally began to calm down. Eventually, he accepted some bananas and seemed to relax.

What made Charley attack Jimmy? We think Jimmy was Charley's first target because he was a stranger whom Charley barely knew, something we'd never even considered when putting all the chimps together. But Charley wasn't finished. In an attempt to stake out this new territory, he systematically attacked and threatened each of the males in the enclosure, much as a dominant male chimp would do in the wild. During the time I was getting Jimmy into the cage, Charley changed his aim and headed for Sandy. Sandy must have walked on air, because he responded by going straight up the front of the cages and out of the enclosure. After we got Sandy back into the cages, we decided to let Tobar in and see if he could comfort his friend Jimmy. Then Rita came up and begged to be let in, stretching an arm out through the wire and looking me straight in the eyes in a pleading manner. So we let her in as well. But out in the enclosure, Charley was still on the rampage, and started after Tara, who got so scared that he fell from a tree at a very dizzying height. Charley grabbed him and jumped on him, but Tara managed to get away and raced to the cages—so we let him in as well.

Next up was Spencer, whom Charley rushed at even though Spencer was in a crouched position with his hand outstretched in a submissive gesture. Charley knocked him sideways, then leap-frogged over Spencer and gave him a backwards kick, wheeling and heading toward Boo Boo, who was sitting atop a tall tree and realized he had no way to escape. He began getting very nervous and started to cry. But as the top branches were not strong enough to support Charley's weight, he shook them violently instead, forcing Boo Boo to try to slip past him on his way down. Charley lunged at him as he passed, knocking Boo Boo off balance, and he fell a frightening distance to the ground, then ran into the undergrowth, with Charley in hot pursuit. Charley caught up with Boo Boo and jumped on him, only giving way when Donna and Cora raced up and threatened him. Chiquito, meanwhile, realized he was next on the list and promptly climbed over the wall and out of the enclosure and retreated back toward the house.

The only male chimp Charley did not attack was Pal, who quickly squatted down and held his hand out to him every time Charley approached. Three or four times, Charley slowed down and seemed to greet Pal, but did not touch him and subsequently raced off to continue his reign of terror. Finally, after more than an hour of chasing and biting and harassing the other chimps, Charley began to settle down. And when we think back on it, he was really only doing what came naturally. Without older role models, chimps—including dominant males—do not know how to behave, and poor Charley was clearly lacking in the etiquette department. He was like a big adolescent trying to be boss, with nobody to tell him when he had gone too far. He would soon settle into the job, though, and as long as the other chimps treated him with respect, he remained calm. We were especially pleased to see his reaction shortly thereafter, when Girly was being very nasty to Big Jane and Charley came to her assistance, chasing Girly away and giving Big Jane a cuddle. It showed that he was learning how to keep law and order.

The next day's official opening proved to be a complete success. We had most of the local dignitaries as witnesses, and Zambia's Permanent Secretary for Tourism and Wildlife came all the way from Lusaka to perform the opening ceremony. He made a lovely speech, and surprised a lot of people by declaring that chimpanzees had rights, just like humans. He observed that Zambia already had quite a few camps for human refugees from war-torn countries and that Chimfunshi was another such camp, only for chimpanzees, whom the poachers had declared war on. There was food and music and a great deal of drinking, and everybody agreed that it was a wonderful day for chimps.

Long after the crowds had left, Dave and I walked up the viewing platform atop the cages to gaze out over the seven-acre enclosure, just to take another look at our chimps sitting in the trees, playing on the anthills, and generally behaving as chimps should behave. It wasn't total freedom—we knew that—but it was a step in the right direction. Dave turned to me and waved his arm out over the enclosure. "This," he said with a smile, "is the culmination of six years' work."

Just then, Chiquito came over The Wall again.

Six

Rita

Rita in 1991

I am convinced that every chimpanzee who comes to us at Chimfunshi has the potential to be happy and healthy, no matter how bad off he or she is upon arrival. But Rita was different. From the start, I thought she was terribly damaged, not so much physically as mentally, and we all believed she had to be retarded. We'd never seen a chimpanzee with so much energy and emotion, yet so little ability to connect to the outside world; she seemed all bottled up inside. I had serious doubts as to whether she could ever be put right.

What happened to Rita was typical of what happens to a lot of baby chimps purchased by childless couples, then lavished with love and affection as if they were human babies. Captured by a poacher in Zaire, Rita was bought by a young couple in Zambia and spent almost two years being treated like their child, smothered with attention and care, and there's no doubt she had the run of the house. She wore diapers, ate from a plate, played with dolls, and was cuddled all day long by her owners—a scenario that *sounds* ideal no matter what the species. Unfortunately, however, the novelty of raising a baby chimpanzee eventually wore off on the husband and wife, and as Rita grew bigger and stronger, she became less and less cuddly. Her owners began to work all day, and I understand that the housekeeper who looked after her soon became the only person able to handle her at all. Rita was kept in a diaper around the clock to avoid making messes, and was eventually consigned to a cage. With no other chimps around to vocalize with and only minimal human contact, Rita's health began to deteriorate, both mentally and physically, and by the time she arrived at Chimfunshi in April 1986, Rita, then just two and a half years old, exhibited all of the spark of a cabbage.

The first thing we did was cut off Rita's filthy diaper, which reeked of excrement and had begun to attract flies. We left her loose on the lawn for a moment and she started to walk around bipedally, like a human, placing her hands between her legs where her diaper had been, as if frightened she had lost something. The look on her face was clearly one of distress. We offered her some fruit—oranges, bananas, and some pineapple; nothing too exotic—but after politely taking each piece, she held it in her hands, examined the object, then slowly put it on the ground and refused to eat. She acted like she'd never seen whole fruit before. But the most awful thing was her rocking. I have heard that human beings suffering from brain diseases such as autism will rock themselves into a trance, repeating the same movements over and over, simply because the repetition comforts them when they become excited or confused. That was Rita—she would sit in a corner and rock back and forth for hours, bashing her shoulders and head against the wall until her skin was raw, and if you tried to get her attention, she looked back with eyes that were vacant and dull. At night, we could hear her the solid *thump, thump, thump* of Rita's back hitting the concrete as she tried desperately to rock herself to sleep, and some nights it lasted until well into the early hours of the morning. I know, because I often sat up listening, hoping each *thump* would be the last, but it was as if she had no control over herself. It was awful.

Because of her bizarre behavior, we thought it best not to put Rita in with the other Youngsters too soon. It took over four weeks for her to relax enough to stop that hideous rocking during the daylight hours—the nights were another matter, unfortunately—and her initial contact with the others was stiff and uncomfortable. Clearly, Rita had no idea what to make of the other chimps, and she surely did not know what was expected of her. But just as chimps have an amazing ability to recover from their wounds, they also have the power to help one another heal, and it wasn't long after Rita had joined the others full-time that she began to make dramatic improvement. It was as if The Youngsters were reaching out to her. Girly in particular formed a sisterly bond with Rita, perhaps passing on much of the peace and security she had experienced after being adopted by Liza, and Bella also became close. Rita's emotional problems began to fade away, and soon she was behaving just like the other chimps. It was about three months after she arrived that I let her out of the cage one day and she turned a double-somersault across the lawn, then loped off after the others with a big grin on her face. I turned to Dave and said, "I think we've got it. She's going to be OK."

It was true. Rita was soon somersaulting all over the lawn, splashing in puddles, fighting over juicy pieces of pineapple, throwing herself onto unsuspecting persons from tree branches, and sleeping in huddles with the other chimps in the same bed, not far from the corners where she used to rock.

As Rita progressed, two things began to happen. First, she began gaining weight at an incredible rate, which led us to wonder if perhaps she wasn't older than we thought—maybe three or four—and had only been small when she arrived because she was undernourished. Rita seemed to grow longer instead of fatter, and her arms were extremely strong for her age. Sometimes, when I bent down over her, Rita would throw her hands over my head and pull me on top of her, then convulse into fits of laughter. She did this because she loved being tickled and knew it was a good way to get what she wanted. Rita would begin laughing even before I touched her—that hoarse "*heh-heh-heh*" pant that all chimps make—just like an infant who knows what lies ahead. Of course, all the others wanted to join in, so even when just Rita and I began the huddle, I usually wound up with a half dozen or more chimps draped across my back, with all of us laughing hysterically.

Second, Rita began to display uncanny maternal instincts. I've always felt that some chimps are natural-born mothers, possessed of the ability to care for others whether they are related or not, and this was surely the case with Rita. Within four months of her arrival at Chimfunshi, she was attempting to cuddle and carry some of the younger chimps, and she even tried to carry Little Jane—in the same manner a mother would carry an infant, slung under her stomach—even though Little Jane was practically the same size as she was. One day, long after Rita had ceased that terrible rocking, I looked up and saw her swaying back and forth in her cage just the way she did when she first arrived, and all I could think was, "Dear God, here we go again." I raced toward her cage, but as I got closer, I saw that Rita was holding Little Jane in her lap and was gently rocking her to sleep. Rita's upbringing—for all its trauma of separation and loss—had somehow left her with a solid foundation of love, and she soon began adopting other little ones, such as Coco and Jimmy, and would defend them without fail in a fight. If two chimps got into a tussle, Rita always came to the aid of the smaller one—and woe betide anybody who dared threaten the littlest babies. Rita took it upon herself to protect the rest of The Youngsters, and sometimes even challenged Charley when he behaved rudely.

Rita's sense of caring extended to the older chimps as well, and she would fuss over anybody that fell ill, peering into their eyes, examining their noses, and showing genuine concern. I used to joke that she was the first full-time nurse we had at Chimfunshi, but it wasn't far from the truth. One morning, I noticed she was engrossed in looking at Tara's foot. Tara was sitting very still and relaxed, just watching as Rita squeezed, then scratched, then squeezed again at a spot on his foot. I called to Patrick and said, "I think Tara must have a splinter in his foot. You must come and see what Rita is doing." Patrick walked over, but Rita had looked up at me when I spoke, and, as though answering me, got a good grip on Tara's foot and pushed it through the metal bars at me, a gesture that clearly meant I should have a look at it. There were indeed two spots on Tara's foot, one where I think a splinter had already come out and another that appeared to still have a sliver inside. First I had a go at getting it out, then Patrick, then Rita again, and then she pushed the foot back to me for another turn. I don't think any of us got the splinter out, but I put some ointment on it and promised Rita I would check on Tara later. She seemed satisfied.

Another day, Donna pitched up with a large thorn in her foot, causing her to stop every few steps and hold the foot in pain. Sure enough, Rita was first on the job. She squeezed and poked and scratched at the spot, sometimes causing Donna to wince so that Rita was compelled to place a hand on her for assurance as she prodded the sore area. Rita only stopped ministering to let me try, but I could not remove the thorn either. When I checked back later in the day, however, the thorn was gone and Donna was acting as though nothing had happened. We were sure Rita had been the one to remove it, especially since she was observed looking again at Donna's foot a few hours later—a follow-up visit, I suppose.

While thorns and scratches and small cuts were commonplace among Rita's patients, she was always willing to let me consult. And she particularly liked when I gave out medicine. Because chimps are so susceptible to colds and flu, we have always been quick to counter the slightest hint of colds at Chimfunshi with cough medicine and, when necessary, antibiotics. The chimps loved the syrupy-sweet taste of cough remedies, and we'd sometimes mix the bitter pills with honey to disguise their taste. Rita was fascinated with the whole process of medication, and would follow me from chimp to chimp as I gave each their dose, then demand hers. But because she seldom got as sick as the others—and recovered more quickly when she did fall ill—I often left Rita until last, causing her to get quite upset if she thought she was missing out. Rita was particularly engrossed the time Cleo got bitten on the foot by a snake, a wound that became quite swollen and painful and forced me to clean the spot every day. But just getting at the foot was a difficult task, as Rita's furry black head was always in the way, peering intently at the spot and licking off any ointment or salve I tried to apply.

I was soon playing doctor and nurse to far more than chimps at Chimfunshi. Knowing that we could never afford to purchase all of the antibiotics and pills we needed for the chimps, many of our supporters began donating medicine—most of it outdated human remedies that were still effective on chimps—and I managed to amass a small pharmacy in just a few months' time. In a country like Zambia, where health care is either too expensive or inconvenient to be effective for the general public, word of the availability of medical aid spreads quickly, and Chimfunshi soon became a regular field hospital. Many was the morning I'd look out and see a line of locals—from mothers with sick children to ailing old men and women—waiting patiently for me to come look them over, and my nursing skills were often put to the test.

Fortunately, we had a wonderful veterinarian staying with us then, a young Frenchman named Philippe Bussi, who pitched in and helped with a lot of the cases—both human and chimp. Philippe was a bit eccentric, and he wore thick granny glasses and a gold earring under his mop of curly blond hair. But he was passionately devoted to chimps, and came to Chimfunshi after spending some time at a chimpanzee rehabilitation project in Gabon. Phillipe stayed for many months, and whether wrapping broken arms in plaster or diagnosing illnesses like malaria and pneumonia, he always worked with a confidence and precision that was a joy to watch.

The availability of health care formed a bond between Chimfunshi and the surrounding villages, one that would grow stronger when we realized we could not produce enough fruits and vegetables on our farm to keep all the chimps fed. So we offered to buy produce from the neighboring farms along the Kafue River, establishing a trade that made the locals interested in Chimfunshi's success. We also purchased sacks of wild bush fruit collected from the forests. Each Monday, a long line of vendors would assemble outside the main gate to the compound, many of them having walked for hours with these large bundles to sell, and we bought everything from sweet potatoes and guavas to cabbages and peanuts, using a large hanging scale to weigh each load. Of course, every transaction required a civil amount of haggling and debate and drew the whole process out, sometimes over the entire morning, but it was all friendly and the truth is we couldn't have kept going without the vendors.

But the Kafue River itself soon became a dangerous point of contention between ourselves and some groups of local fishermen. For months, large gangs had been illegally stringing nets across the river and driving the fish into nets by beating the water upstream, a process that is referred to as *kutumpula*. They were backed by Chanda Chaiwa, a powerful and evil man who lived in a village on the other side of the river. The fishing gangs were breaking the Zambian law that made it illegal to fish during the spawning season, and many of the gangs were armed and belligerent. Since they were seriously depleting the fish in the river—and were, we believe, responsible for a rash of crime on our property—Dave took to confiscating their nets and turning them over to the police. He even hired a group of workers to patrol our banks, and got into several nasty shouting matches with the gangs. They threatened to beat me and our son, Tony, and vowed to kill our cattle. When we found a cow hanging from a tree one day with all the meat removed, we realized these men meant business.

But all our attempts at getting the police or the district governor to help were futile. Zambia's law enforcement was terribly haphazard and corrupt then, especially in an area as remote as Chimfunshi, and Chaiwa curried favor by giving some of the police food and alcohol. We were forced to deal with this problem ourselves. But a few days after a heated run-in with Chaiwa down by the river, Dave was met by a group of policemen, who arrested him and took him by boat to Chaiwa's house. There, Dave was interrogated and verbally abused by Chaiwa's friends, even as the police lounged around outside, then was transferred to the police station in nearby Chililabombwe. They took Dave's shoes and socks and glasses, then forced him to sign some sort of confession. He at first resisted, because he could not properly read without his glasses, but eventually he gave in. Dave was placed in a cold, wet cell where he sat on the floor with a group of other prisoners for ten hours, and we really did wonder how we'd get him out. I was terribly scared and worried about Dave, and none of the police officials I could find were willing to help. Fortunately, I was able to convince an army colonel to intervene, and he ordered Dave to be released, reprimanding the police officers in the process.

At times like that we wondered why we stayed in Zambia and put up with so much, especially the rampant lawlessness. But we figured those sorts of things happened the world over and that it would do us no good to run away. We just resolved to do the best we could.

Dave's terrifying ordeal did have a strange ending, however. When Dave was arrested, I got so angry that I hastily drew a caricature of the district governor and put an arrow through it, telling our housekeeper, Patrick Katwamba, "Nobody does that to my husband and gets away with it." A few months later, all of our staff showed up as usual one morning but refused to work, standing about sullenly outside the house. I asked them why they were not doing their jobs, and one of them told me the district governor had died the night before from my "curse" and they were afraid of me. I did my best to keep a straight face and hide my feelings. The district governor was known to have been quite sick for some time and his death was not unexpected.

A year later, the men again came to work one day and just stood outside the gate, refusing to enter the compound. I went to see what was wrong, and was told this time that Chaiwa had been found dead the night before, just down the river from our house. Once again, my "curse" had been successful—though the truth was that he disturbed a female hippopotamus with a baby while putting out his nets and was crushed trying to get away.

Meanwhile, Rita was proving to be one of the most intelligent chimpanzees at Chimfunshi. Like many chimps, she was obsessed with shoelaces, and wasted little time in removing them from the shoes of guests, who did not seem to mind. But whereas most chimps would disappear up a tree with the string and slowly rip it to shreds, Rita, with great concentration, would actually attempt to relace the shoe. She never seemed to get it through all the right eyes, but she did usually manage to get both ends of the lace into some eyes, and would then try to tie it up again. It was not quite the job you would do yourself, but for a four-year-old chimp, it was pretty good.

We once had a young American scientist named Mark Wright stay with us for an extended period, and he regularly accompanied the chimps out on their daily walks in the months before The Wall was completed. One day, Mark said that it looked like rain, so he put on his raincoat, which hung down to below his knees. Before the rain actually started, however, Rita got under the coat and fastened herself to Mark's legs, then stayed there safe and dry as the rain came down. And when Mark attempted to walk away, Rita held tightly to his legs for a short distance, then dashed off and found shelter under some bushes. Mark found it fascinating that Rita was clever enough to anticipate rain, either because she'd seen the clouds forming or because she'd seen him put on his raincoat. Either way, it was clear she understood what the raincoat was for. Not long after, Mark began allowing Rita to use his notepad and pen to make drawings out in the forest. Her scribbles were very interesting, because they seemed to have a pattern. She repeatedly made inch-long ticks ending off with a long curve, a bit like a funny D not joined at the bottom, and would draw as many as seven in a row. She also made a drawing that looked amazingly like a flower, with the petals out around it. Mark found it difficult to let her draw with his pen and paper when the other chimps were around, though, since each of them wanted nothing more than to annoy Rita by stealing her artwork. He eventually was forced to wait until they were napping, around midday, then call Rita over and quietly slip her the tools. She'd sometimes sit and draw for thirty minutes or more.

Rita proved to be a fast learner in the forest, something that was a great relief to us because we'd feared that so many years in a human household might have dulled her natural instincts. She was a hopeless nest-maker at first, but after watching Boo Boo closely once or twice, she began to make expert nests, well situated and strong enough to hold several chimps. Rita's appearance also continued to change at an astounding pace. Although she could always be identified by an old wound on her right ear—a walnut-sized hole shaped like a half-moon—Rita's body shape quickly went from long and leggy to full and stocky, with a beautiful thick coat that was very shiny. She was really quite beautiful, and Dave took to calling her Sexy Rita. He'd walk up to her and say, "How's my Sexy Rita?" and she would practically throw herself into his arms for a hug.

Rita's creativity and ability to amuse herself meant she was always inventing games. One day, when I gave the chimps a large pile of oranges at their midday meal, Rita collected about fifteen of them into a neat circle on the floor, then put her arms around them and started moving backwards, dragging the oranges with her. She was in a very happy mood and seemed to be laughing to herself. She went backwards and forward around the cage like this for a long time, stopping every now and then to do a headstand on the oranges or to jump on them, scatter them, and then collect them all in a pile again. She then got one of the tires from the sleeping area, put the oranges inside it, tipped the tire up on its rim, and started rolling it around the cage, all the while watching the oranges as they slowly tumbled around and around inside the tire as if they were in a clothes dryer. She let go of the tire at one point and Jimmy, who had been watching her with keen interest, rushed over and knocked it down. She got very angry with him and started to chase him across the cage. While she was chasing Jimmy, Sandy went and put his head through the center of the tire and seemed about to steal one of her oranges. But Rita saw and shouted at him, and a small fight broke out. Sandy and Rita rolled round and round in the hay, screaming and flailing until the others came and separated them. Rita went back to her oranges and proceeded to eat five of them, while the other chimps grabbed up all the rest. This game lasted—with some modifications—for six straight days at lunch, although it did not always end in a fight.

Another of Rita's passions was cleaning her cage. She began by studying our routine very carefully, noting that we would first remove the chimps' straw bedding, then sweep the floor. Once a week, the walls and floors would get scrubbed and be left to dry before fresh straw was placed into the cages around midday. I was shocked the first time I saw her collect a pile of straw in her hand, then carefully pick at it and twist it until it resembled the large grass brushes we used to clean the cages. Rita would use the brush to sweep the floor of her cage, collecting all the old straw and fruit skins and so on in a big pile in the middle of the room, then make a big sweeping motion and scatter all of the rubbish about the cage. And then she'd start again, cleaning and scattering, cleaning and scattering. And when not cleaning the floor, Rita would pour water on the cement table in her cage and use the straw as a sponge to scrub down the top until all of the water was gone. If she ran out of water, Rita would hold out her hands and beg for more till she felt she'd gotten the table as clean as she could.

Rita became obsessed with a wide variety of water games. She began by cooling off after each day's bush walk by running over to the drinking trough and sticking her head all the way under, then lifting it up and laughing crazily as the water ran everywhere. It was probably her way of taking a shower, but she did look quite funny, and the other chimps would gather round and point and laugh. At other times, Rita would make a sponge out of bits of grass or lettuce leaves, then get this wet and scrub her hands until it disintegrated. She would repeat this process for up to fifteen minutes, hunt around for any object she could fashion into a cup—such as an avocado skin or an orange peel—and spend hours pouring water all over herself. And if we'd encountered heavy rains and the chimps happened upon a mud puddle anywhere on their bush walks, you could be sure Rita would quickly throw herself into it and roll around, then lie on her back as the others raced up and jumped in, causing huge splashes.

When we moved the chimps into the seven-acre enclosure in 1989, Rita at first acted the perfect den mother, looking out for the little ones like Jimmy and Cora and keeping them clear of any disagreements between the older chimps. In fact, Rita always seemed timid around the big chimps, especially Charley, and preferred to stay high up in the trees whenever he and the elder females like Liza were acting edgy. At night, she usually chose to sleep inside the cages with the other Youngsters, often cuddling up close to her best friend, Tara, or Jimmy. Yet there were times when Rita herself acted like a child, and there's no denying she was growing increasingly fascinated by Charley—even if from a discreet distance. Sometimes I'd catch her staring at him as he ate, or inching closer to him as he lolled in the sun, even though the fear in her face was plain to see. Once, after Charley had been playing with The Youngsters for over half an hour—chasing them around the bushes, rolling on the ground with them, and tickling them—Rita grabbed his hand and began running round and round him as though he were swinging her. Both their mouths hung half open in play-grins, and their laughter was audible from far away. His attention seemed to mean so much to her.

But Rita soon found that Charley's attention came at a high price. Most female chimps come into oestrus at about the age of nine or ten, a condition that includes a swollen pink bottom and a bloody discharge that signals to the males a female's sexual maturity. Rita began showing signs of coming into season in early 1992. She and Donna were on the same cycle, and neither appeared pleased with her condition. They kept well away from the big males, and on occasion, they refused to come for their morning milk because the boys had been hanging around the cages a little too long. Rita in particular seemed to be asking more and more for a little cuddle or a bit of handholding, as though in need of some reassurance, while Donna simply sat in the cages with her arms around her stomach, hugging herself and wanting nothing to do with me.

It did not take long for Charley to notice all this, and he quickly decided that Rita was to be his special friend. He insisted that she stay near him, and he spent days chasing all the other males away. Rita decided, however, that she wanted none of Charley and kept running away from him, often hiding in the bushes or cowering inside the cages, since Charley was too large to fit through the door. Charley was furious! He was angry with everybody because he was not getting his own way, and after two days of playing hide-and-seek, Rita finally consented to stay with him during one lunch hour. He thought he had won, but as soon as lunch was over, Rita managed to get herself lost in the bushes again. Poor Charley was so incensed that he sat on the ground and cried in sheer frustration.

This battle of wills between Rita and Charley went on for months and months. Every time she came into oestrus, Charley would indicate he wanted her to stay close by. But Rita would automatically go the other way, and I wondered after awhile if perhaps she wasn't enjoying the game. Clearly, she would not conform and do what Charley wanted, nor did she go off with any other male, either.

Her flirtations did ignite the first power struggle we'd ever had at Chimfunshi, however. Pal had made a few halfhearted attempts at challenging Charley, but he lacked the support of the dominant females that is crucial for success. His easygoing nature may have been to blame, or perhaps his disfigured face. But I've also felt that truly dominant chimps are born into the role, and maybe Pal's family in the wild hadn't been very powerful. Still, because his periodic fits did terrify the other chimps, he was able to become Charley's confidant instead. And when Pal stepped aside as a threat, Spencer quickly filled the breach.

Spencer had watched Charley's struggle with Rita closely, and I think maybe he'd gotten some big ideas. All I know is that Rita was at her most fertile on August 20, 1992, and that was when the trouble started. As I gave the chimps their milk, I noticed that Charley was impatiently indicating to Rita to follow him. Spencer, meanwhile, was sitting quite close to the cages, keeping a careful eye on Charley, but also motioning to Rita to join him every time Charley turned his back. Caught in the middle, Rita kept grinning nervously, but eventually followed Spencer into his cage to drink her milk. Though the door of Spencer's cage opened sufficiently to let him in, a chimp as broad as Charley could not squeeze through, and when he realized where Rita was, and with whom, he went berserk. He rammed against the steel door and kicked it repeatedly with resounding thuds, a noise so loud that Rita thought she had better give in, so she jumped through the door and followed Charley into his cage. Everything went quiet, so I left them until lunch.

I did not see what happened next, but Patrick later told me that Spencer had started the fight. I was getting food ready when I heard a terrible screaming of a lot of chimps, and by the time I got to the scene, Spencer was just pulling away from

Charley and Liza Do Little, his face covered in blood. Patrick said that Spencer had run at Charley on two feet, his hands held high and shaking. The chimps had seemed to connect, and suddenly there was a big ball of heaving black muscle rolling around, to the accompaniment of loud, raucous screaming from the rest of the chimps. But a fight between two male chimps—especially if one is the dominant male—does not stop there. The dominant females often join in to defend their male, which is why Liza had suddenly crashed through the bushes and joined the melee, grabbing and biting at Spencer. At one point, her hand streaked across Spencer's face, ripping it open in a number of places, and that's where the fight ended.

We managed to get Spencer away from the other very curious chimps, and put him into a cage. His top lip was ripped in two, right up to his nose, and his top gums and teeth were all exposed. There was still blood everywhere, and Spencer's lip was swelling fast. My poor handsome Spencer did not look so handsome anymore.

Interestingly, this was one time Rita did not help treat another chimp's wounds. In fact, she was nowhere to be seen. Clearly, she felt she'd already done enough.

Seven

Sandy

Sandy in 1988—up to mischief, no doubt

In any children's classroom, you get the naughty bad boys and the naughty good boys—and it's always the naughty *good* boys who steal your heart. Well, that was Sandy, an incredibly mischievous chimp who somehow managed to make us love him. He stole my heart almost immediately, and even though his antics and endless escapes eventually made him an outcast from the others, I'll never think of him as anything but a lovely, handsome chimp.

It was early 1988 when a game ranger informed us that someone in Kitwe had two baby chimps that were to be used as an attraction at an automobile repair shop. The game rangers were waiting to move in to confiscate the animals until they could be sure we'd accept them. Of course, we said yes, but we also wanted to try and handle the situation as delicately as possible.

Owning chimps was against the law, but we still found it hard to confiscate these two, because the person who had the chimps appeared not to have known the law when he bought them. For that reason, we invited the owner to come out to Chimfunshi, have a look at our facilities, and hear what we were trying to do. We thought maybe we could convince him that surrendering the chimps to us would be the best thing for all concerned. Unfortunately, the owner was very upset when he found out it was illegal to own the chimps, since he had paid a lot of money for them, and he took great pride in describing his big plans for a very small cage he intended to build to draw crowds to his garage. The fact that the chimps would be poked and teased and tormented seemed immaterial to him, and he clearly had very little feeling for them. But after finding that he could not secure a permit to get them out of the country to try and sell them elsewhere and recoup some of his money, the man eventually brought the chimps to us.

Sandy, an eighteen-month-old infant male, and Cora, a slightly older female, arrived together in March 1986. Sandy was quite small when he came, weighing just fifteen pounds, and he seemed possessed of a permanent cold and a perpetually runny nose. He was always a mess, and I took to stuffing large amounts of toilet paper in my pockets whenever I went near him for wiping his nose. At first, we thought Sandy was the insecure one of the pair, since he always seemed to be running to Cora for comfort and was usually in a pout, but later we realized that it was Cora who needed someone to cling to, especially when she was upset. In fact, Sandy eventually made it clear that he did not need to be hugged too often, pushing her away and wriggling out of her grip, and his personality—that of a snooty-nosed little extrovert—quickly emerged.

Sandy figured out the odds faster than any chimp I have ever seen, and soon had all of us scrambling to keep up. When we had visitors, Sandy would hang on to the wire of his cage with a look of such complete innocence on his face that everyone would be deceived. They'd get closer to stroke the lonely little orphan, so lost and unloved in a cruel world, and suddenly lose the glasses or pens they'd had in their top pockets—

and off would go Sandy with a laugh on his face, clutching his new toy and turning cartwheels. He knew this sort of act would bring the house down, and soon we'd be chasing him here and there to try and retrieve whatever he'd stolen, but Sandy would always dance just out of reach, his chest heaving in hoarse laughter. He clearly enjoyed the calamity, and never tired of thinking up new ways to cause trouble.

One day, a visiting family included a little girl who wore glasses with thick plastic lenses. We warned them about Sandy, of course, but it wasn't long before he was in full pout, his lower lip stuck out as far as it would go, and the little girl was stroking his right hand and sweet-talking him through the wire. Suddenly, Sandy's left hand streaked out and snatched away her glasses, all in one lightning-fast move, and he began his usual victory celebration. But for some reason, Sandy stopped and took particular notice of these glasses, and the fact that they were a little dirty. He peered intently at the lenses as he held them up to the light, rubbed the plastic with his fingers, then decided to clean them by rubbing the lenses on the concrete floor of his cage. Needless to say, the glasses were ruined. I apologized profusely to the family and made sure the little girl was all right, and they made me feel at least a little better by telling me she had an extra pair at home. As they left, I overheard the little girl giggling and asking her mother when they could visit again.

That encounter forced me to start giving our guests a good once-over when they arrived, and I'd ask them to remove all necklaces, earrings, glasses, pens, and anything else that The Youngsters could grab. I forgot to tell one man to remove his baseball cap, though, and it lasted all of about two minutes. Sandy held his hand out in greeting, the man leaned forward to shake hands with a look of complete delight upon his face, Sandy's other hand shot out, and the hat was gone. The chimps all thought the hat was a particularly great prize, but in the battle for possession, it got torn into about six different pieces. Sandy ended up with the brim, which he held on to for the rest of the day, alternately using it to shade his eyes and brush his teeth.

Once he had mastered the art of stealing, Sandy began a second hobby: throwing things. Whenever he'd finish eating an orange or a banana, Sandy would gather up the peels and throw them at anyone within range—usually with stunning accuracy. It wasn't long before mealtimes became a regular battlefield with Sandy around, as bits of fruits and vegetables filled the skies like V-2 rockets; rare were the times I emerged unscathed. Once, Sandy hit a gentleman with a maize ball in the side of the head, and it exploded in a cloud of white flecks, leaving soft pieces of maize stuck to the man's hair and face while others tumbled down into the collar of his shirt. Another time, Sandy scored a direct hit on a woman's ample bottom with a large orange, and the resulting *smack!* was so loud that it caused me to spin around and see who had clapped their hands. Sandy eventually became a connoisseur of throwable food, preferring more solid bits like apple cores or apricot pits, or fruits that had peels he could wad into a tight ball. He clearly eschewed leafy foods like lettuce and cabbage, though. The few times Sandy threw those, the leaves just fluttered harmlessly to the ground, and he trudged away, disgusted.

Sandy threw his food because he knew it drew attention to him and because we usually offered him something else to eat if he'd stop. But I had to laugh the day Sandy spied a particularly good target and instinctively threw a half-eaten orange. Then, realizing he'd made a mistake, Sandy held out his hand and begged for the fruit back. He retrieved the orange, finished eating it, then threw the peel at somebody else.

But if Sandy enjoyed antagonizing the visitors who came to Chimfunshi, his real joy lay in terrorizing the other chimps. Sandy's favorite game was to sneak up behind a chimp who was sitting quietly, then belt him or her in the back of the head and rush off, expecting to be chased. Or, if Rita or one of the others was playing with something, Sandy would wait until they were looking the other way, then go and steal their toy—again, causing a great chase to ensue. We used to give the chimps fresh sugarcane, cutting the stalks into large, two-foot pieces, but were forced to stop when we realized all we were doing was arming Sandy. He would invariably start fights with the stalks of cane, hitting the other chimps on the head and shoulders, and then do half-cartwheels while getting out of their way. The other chimps usually ganged up on Sandy and noisily chased him off—but he never learned. Sandy once lost his piece of sugarcane during one of these melees, and came rushing up to us to beg for some more. But when we told him it was all finished, he lost complete interest in the chaos he'd initiated and just sulked for a while.

Despite his outrageous behavior, Sandy made friends surprisingly quickly. Rita became a confidante, and Cora remained something of an older sister for a time, but most of Sandy's time was spent cavorting with Tara. They were thick as thieves, and both seemed to get such fun out of life, scrambling up trees or picking play-fights with the other chimps. Sandy and Tara appeared to have memorized the location of every tree and bush in the forest, and would rush off ahead and either sit in the low branches to pounce on visitors as they walked past or pull back the saplings like catapults, releasing them just in time to slap an unsuspecting visitor in the face. Other times they would hurry ahead, then quietly double back and slip up from behind, diving onto your head and shoulders when you least expected it. Few guests returned from those bush walks without a bruise or two and a torn piece of clothing, but none ever appeared angry, much to my astonishment.

Sandy and Tara got the fright of their lives one morning, though, when a young baboon got loose from his cage and spent about three days scampering freely through the main compound. He was no trouble really, except when the chimps came out of their cages to go into the forest, and one particular morning, Sandy and Tara decided that nobody was going to have that much fun without including them. They wriggled out of our arms as we were carrying them out and took off after the young baboon, chasing him all over the garden, but the baboon was fast and nimble enough to evade them. And when the baboon ran over to the older baboons' cage and climbed onto the roof, Sandy and Tara were in such hot pursuit that they followed without stopping to consider where they were going. Immediately, three very large male baboons rushed at the chimps with a roar and grabbed at them through the wire, and Sandy and Tara were so frightened that they both let go, dropped about ten feet to the ground, and ran screaming past the house out the main gate and into the forest after the other chimps. When they found them, Sandy and Tara sat on the ground and hugged each other for a long time, panting hard. They never went near the baboons' cages again.

Sandy and Tara were as impish as they were unpredictable, and never failed to seize an opening. When Carole Noon made her first visit to Chimfunshi in 1989, she regularly accompanied the chimps on their bush walks, keeping careful notes and helping us to better understand chimpanzee behavior. One morning, Carole apparently consumed too much coffee before heading out, and found herself needing to use the bathroom au naturel. If slipping away from a mob of eleven young chimpanzees proved difficult, however, avoiding Sandy and Tara as she attempted to remove her trousers proved virtually impossible, as Carole later wrote in a letter:

> I waited until Sandy and Tara were up a tree and quietly snuck away—or so I thought. Hidden behind an anthill, my pants around my ankles, I heard someone coming. I started to stand, and in an instant, Sandy and Tara were on me. I was, of course, at a disadvantage what with trying to pull my pants up and all.
>
> I was doing a fair job of fighting them off when I lost my balance and fell to the ground, my pants around my knees. Suddenly, it became Tara's life's ambition to possess the hot pink underpants that I was wearing. Now, in my own defense, I would never go out and buy hot pink underpants, (but) these came in a package of assorted colors, you know, five for six dollars or something like that. Tara ripped one entire side of the underpants. The sound delighted Sandy and he appeared to understand Tara's goal and set about helping him. They both started pulling. What was I to do? With my pants around my knees, was I supposed to call for help? I heard more ripping, and now the other side gave way. They were close to getting them off me.
>
> I know this sounds silly, but, somehow, if the underpants had been white, or black, or even blue, I wouldn't have been so upset. But I had a picture in my head of Tara and Sandy running off, waving my hot pink

> underpants through the bush. And, of course, Patrick would be hot on their trail, yelling 'Bring Madame back her panties!' Curious, the others would get involved. In the end, I knew they would rip the pink underpants in little pieces and each carry a hot pink patch home with them. With this image in my head, I mustered superhuman strength, and, hanging on by a thread—literally—I won this encounter.

For all his devilment and deceit, Sandy's intelligence was what impressed us the most. Chimps who have been orphaned at a young age are invariably slower to learn the proper skills, since chimpanzees usually develop by watching and imitating their mothers or older siblings. But Sandy quickly overcame such handicaps. When it came time to nap, for instance, Sandy was too little to make his own tree nests, but he watched Cora closely when she'd make one up, and it wasn't long before he knew all he needed, carefully bending the tree branches and leaves into a loosely braided circle, then using his body weight to hold the nest together.

Sandy's skills of observation extended to humans as well. On a bush walk one day, Patrick was suffering from a head cold and stopped in the forest to sneeze. He then pulled out his handkerchief to wipe his nose—and noticed that Sandy was staring intently at him as he did so. Sandy looked around, then picked up a big green leaf and rubbed his own nose from side to side, all the while continuing to watch Patrick. So Patrick again wiped his nose, and Sandy again wiped his. But before Patrick could wipe his nose a third time, Sandy grew tired of the etiquette. He tossed the leaf aside, quickly wiped his nose with the back of his arm, and scampered off.

At night, when he tried to find a comfortable place to sleep, Sandy often wound up crying and shrieking because someone had stolen his spot inside the sleeping boxes. He'd jump up and down and scream at the top of his lungs until usually Rita or Cora gave in and grabbed hold of him for a quick cuddle. I often wondered if those outbursts were a variation on the nightmares that used to wake Pal each night, or just Sandy's way of getting attention. But one thing was clear: Over time, Sandy's pouts and tantrums became central to his behavior. When we had a severe outbreak of colds a few months after he arrived, all of the chimps were affected—but Sandy somehow seemed the most miserable. He had great difficulty drinking his milk in the mornings because he could not breathe properly through his nose, and I think I must have used up a roll of toilet paper each day on his nose alone. Sandy reverted to being a real baby during those couple of weeks, insisting on being carried everywhere, and at night he'd cry out to be hugged if he thought no one was taking notice of him.

When we took the chimps out on their morning bush walks, Sandy insisted on being carried, and if you dared set him on the ground and tell him to walk, he'd just sit there and shriek until you picked him up again. One morning, I really tired of this behavior, so I put Sandy down and turned my back on him. He pitched a terrific tantrum, throwing himself on to his back on the ground and kicking his legs and flailing his arms about, all the time shrieking loud enough to be heard for miles around. It was one of the most impressive fits I have ever seen a chimp throw. In fact, the rest of the chimps—who normally ignored one another's outbursts—all stopped what they were doing and came over to see what was happening. Tara soon leaned in to Sandy and hugged him, and although Sandy screamed a little while longer, they eventually walked off with their arms around each other. Normally, I'd get cross when Sandy screamed like that, but the fact that he found comfort from another chimp—rather than a human—gave me hope that he might eventually come right.

Sandy's flair for drama kept us on our toes. When he suffered a cut on his right index finger one afternoon, you'd have thought he was mortally wounded. No one knew how the cut had happened, but we suddenly heard him screaming and rushed over to see what was wrong. By the time we got to him, he had stopped screaming and was busy trying to stem the flow of blood. He was licking the wound and sucking it while all the other chimps were trying to get as close as possible to see what was wrong. We eventually stopped the bleeding, then I washed the wound and put some antibiotic cream on it. He immediately licked it off, so I put some more on and made sure it went well into the cut. He licked that off too, then let Tara have a go at licking it. All the others wanted a lick too, but Sandy decided they were only allowed to look. It wasn't long before Sandy appeared to be showing off this cut, as he kept insistently flourishing it under Tara's nose. I have often seen little boys proudly show off wounds to their friends, and I wonder if this wasn't what Sandy was doing. He was the center of attention, and reveling in it. We never did figure out what caused his cut, but it healed without becoming infected, even though anything we put on it was licked off—even gentian violet, which turned his lips purple for a week.

One morning, Sandy seemed a bit mopey. He was sitting with his arms folded, appearing to feel very sorry for himself and refusing his milk. I had a good look at him and discovered that he had a loose tooth. It was a baby tooth that was ready to fall out—there was already another little tooth behind it. His gum was a little bit hot to the touch, and I imagined he was in pain, but Sandy milked the moment for all the drama it was worth. He spent the entire day as if on the verge of death, barely stirring himself when another chimp appeared, seemingly too weak to even stand. But Sandy was back to normal the next morning, and he proudly showed me where the tooth had fallen out by pulling back his lips with his fingers.

When we finally opened The Wall and mixed all the chimps together in the seven-acre enclosure, Sandy was forced to alter his behavior, and he seemed to adjust quickly to the community beyond The Youngsters' social group. Sandy used to terrorize the other chimps on the daily bush walks and at feeding time, but now he suddenly found himself confronted with older chimpanzees that were unlikely to tolerate such behavior. We held our breath, fearing that Sandy's antics would run afoul of Charley or Spencer or one of the other big males. But to our surprise, Sandy seemed to comprehend the politics of the new "order" within The Wall, and regarded the big chimps very warily. Whenever he saw Charley coming, Sandy would dash back into the cage with a look of nervous fear on his face and cower inside, even though he was still brazen enough to stick his head around the door and shout at him as he passed, knowing that the cage door was too small for Charley to get through.

Scoundrel, clown, bully, brat—this was the Sandy we came to know at Chimfunshi, a chimp who arrived as scared and confused as any other, but quickly developed into one of the leaders of The Youngsters. Both Sandy and Tara had begun to show signs of becoming proper male chimps in many ways, rushing to protect or cuddle others when trouble broke out, or stepping in to settle disagreements and fights among the group. They also spent more and more time in the trees together, taking naps or just passing the hours, although they quickly moved off if one of the adult males got too close. Sandy was also growing rapidly in size, and more than once I glanced into the enclosure and had to look twice to make sure it was him. His shoulders and back were developing powerful muscles, and there was no doubt he was going to be an awfully big chimp. His lovely black coat was also particularly beautiful, and it would shimmer brilliantly in the sunlight.

Which is why the events of April 30, 1990 came as such a shock. At six o'clock in the morning we heard an awful screaming coming from The Wall, and we knew something terribly wrong must have happened. Dave and I rushed over to the viewing platform, peered through the morning haze into the enclosure, and saw a few chimps sitting near the handling cages with looks of absolute terror on their faces, each casting nervous glances back toward the river. One by one, we managed to get the chimps inside the cages, with Cora the last to arrive. But one chimp was missing: Sandy. Dave and I were quite worried and were preparing to go in and look for him when suddenly we saw this crumpled black mass, this *thing,* crawling very slowly up the path from the river. It had to be a chimp; it had to be Sandy—but what had happened? He took one or two more half-steps, then slumped down quite a distance away, looking as though he could go no farther. It was clear Sandy would never make it as far as the cages, so Dave went into the enclosure and very carefully carried him out. Sandy was covered in mud and blood and looked such a pathetic sight. He managed to hold on to Dave, but otherwise seemed very weak and ready to pass out.

Dave carried Sandy into the house and we laid him on the kitchen floor, and it was clear his condition was critical. There were long, deep wounds along Sandy's left foot and both legs, and his penis and anus were ripped open by a series of smaller cuts. Sandy's face was a mess, his cheeks and lips bruised and swollen and his right eyebrow badly cut up and so much blood inside the eye itself that you couldn't even tell if the eyeball was still there. His head, meanwhile, had a series of large gashes in it, as if he had been beaten with a blunt object. Sandy must have lost a lot of blood, and was so weak and disoriented from his ordeal that he just lay there before us, moaning weakly and breathing in a soggy, raspy way.

Dave and I couldn't imagine who or what might have done so much damage to Sandy. But there was no time for speculation—he needed medical attention and he needed it fast. Unfortunately, our veterinarian, Philippe Bussi, was away at the annual agricultural show in Kitwe, easily a two-hour drive away. This was the first chance Philippe had had to get off the farm and have a little fun on his own in months, and I knew he'd been eagerly awaiting the chance to go drinking and dancing in Kitwe. Now all I could think of was the fastest way to get him back.

Our son, Charles, came in, and I shouted, "For God's sake, get to Kitwe and find Philippe and bring him home! *Now!*"

Meanwhile, Dave and I did what we could, cleaning Sandy up and trying to assess how bad his wounds were. My nurse's training hadn't prepared me for something like this. It looked as if Sandy had been in an automobile wreck. There were so many places where the flesh was laid open like raw meat that it was difficult to find some of the cuts because of all the blood. Dave and I worked diligently, but there was little we could do. Poor Sandy was smart enough to know that we were attempting to help him, but everything we did seemed to cause him great pain. As the hours went by and Charles failed to return with Philippe, I grew very worried. I strained to listen for any sound that might be Charles returning with Philippe, but as the afternoon passed and night began to fall—making it too dangerous to attempt the roads between Kitwe and Chimfunshi—I accepted the fact that help would not come before the next day. So I sat up with Sandy all night, holding his hand and talking softly to him. He screamed every time I tried to leave him for a few minutes, and it was clear he was in a desperate way.

As dawn broke, we heard the sound of Charles's truck rumbling up the road and pulling into the compound. He skidded to a stop in front of the house, and Philippe seemed angry as he got out.

"Was it really necessary to bring me back?" he asked.

"Just have a look at Sandy," I said.

Philippe entered the kitchen, took one look at the pathetic mass of fur and blood and mud before him, and promptly rolled up his sleeves and dived right in. If it hadn't been for Philippe, I feel certain Sandy would have died that day. He anesthetized the chimp and we placed him on the kitchen table, and Philippe started cleaning, cutting, and stitching, a job that took well over three hours. I assisted as best I could, turning Sandy this way and that, and we worked quickly under the weak light of reading lamps. I could not tell you which part of Sandy looked worst. His eyes probably seemed the most badly injured because they were so visibly damaged, and Philippe was forced to cut away a lot of Sandy's eyebrows and stitch them all together again. He also sewed up the big holes on his head, and closed the wounds as best he could along Sandy's legs and genitals. But there was no way of knowing whether Sandy's vital organs were still in one piece.

That first operation established a pattern that Philippe repeated with Sandy every day for over a week. Each morning, Philippe anesthetized Sandy again and cut away any dead tissue, then cleaned some wounds and re-dressed others. He also took stitches out and put more in, working as quickly as possible so that Sandy needn't be under the anesthetic any longer than necessary. Unfortunately, much of Philippe's work involved repairing the damage wrought by Sandy himself—the chimp systematically removed a lot of the bandages and dressings, and even picked out the stitches that he could reach. But with lots of perseverance, Philippe won out, and at the end of three weeks, Sandy was beginning to look and act like himself again. It never ceases to amaze me how quickly a chimpanzee heals. In almost no time, it became difficult to see where Sandy was hurt, due partly to Philippe's expertise and partly to the recovery power of chimps. There was one place that never correctly healed, though—his scrotum. When Sandy urinated, his urine came out sideways and hit him in the leg, forcing him to adopt some strange poses in order to remain clean when nature called.

Once Sandy was out of danger and resting comfortably under anesthetic that second day, Dave and I entered The Wall and began trying to piece together the puzzle of what had happened. We didn't need to look far. Down by the water, at the river's edge, the telltale signs of a crocodile attack were plain to see—the footprints and tail-markings and churned-up earth. Sandy had been attacked, probably as he'd stooped to drink out of the river; the wounds along his legs were consistent with those of a crocodile's bite. His strength and quickness were probably all that had prevented him from being dragged into the river and drowned. But what about the rest of Sandy's injuries, especially the cuts along his penis and anus, which were much smaller, or the terrible wounds on his face and head?

What we thought happened was this: After the crocodile got him, it tried to drag Sandy into the water, since crocodiles kill their prey most efficiently by rolling them over and over until they drown. But Sandy must have been flailing about, and he must have grabbed on to other chimps, who, terrified in their own right, must have begun fighting with him. So as Sandy tried to fight off the crocodile on one side, he was battling with the chimps on the other. But were the chimps simply lashing out in fear over the sudden crocodile attack? Or were they in some way paying Sandy back for all the fights he'd initiated and tricks he'd pulled over the years? It's impossible to say, but one thing is clear: They left him. I mean, they could have finished Sandy off, for chimps do kill other chimps from time to time in the wild. But in this case, they had a go at Sandy and then walked off.

After the attack, Sandy had no faith in chimps anymore—never mind that he was already one of the biggest or the smartest of the lot. He just didn't like being around chimps, and was terrified to go into the enclosure with the others. He seemed particularly frightened of the big females, and it's clear he feared another attack. We never learned who was truly responsible, but I always felt it was just a few of them. Tara would certainly never have attacked his friend, nor Rita; it wasn't in their natures. We don't know if Cora attempted to help Sandy during the attack, but she was the last one to enter her cage when Dave went into the enclosure to get Sandy out, and during his convalescence she also kept coming and looking into the cages, trying to sneak a peek at Sandy. We do not know if this was out of curiosity or if she had a wish to protect him, but afterward she still acted like his "big sister" from time to time.

When Sandy was removed from the enclosure, Rita and Tara seemed to be the ones most affected by his disappearance. It was hard to tell what they thought, or even if they knew anything, about the attack, but Tara appeared particularly lost. At times, he would look us in the eye and thrust out his hand as though asking for something. Approximately two weeks after the attack, we put Sandy in a cage next to the other Youngsters, and then let Rita and Tara in with him. The greeting between the three of them was amazing: They rushed to each other, all of them whimpering, and got into a big huddle, their arms around one another, touching, patting, kissing, and often just looking. Rita and Tara examined Sandy all over, taking stock of his wounds, and even tried to remove some of the remaining stitches. Tara was particularly concerned with Sandy's eye, sitting as close as possible to his friend and peering into his face, touching his torn eyebrow, peering again, then touching again, almost reverently. We watched, holding our breath in case Tara accidentally hurt Sandy. But Tara was incredibly gentle. And Rita, of course, examined every single stitch and scar. The only time Sandy got a bit worried was when they tried to examine his scrotum, whereupon he moved away and covered himself with his hands. We left them together for about three hours, then opened the door to the enclosure. Tara and Rita both left the cage, then turned and waited for Sandy, but Sandy refused to go. We did this every day for a week until we realized his terror was still too real. Sandy needed a fresh start.

Eight

Milla

Milla was older—and smarter—than we could have imagined.

Milla is without a doubt the most intelligent chimpanzee I have ever known. She came to us a tired, graying old lady with a temper—suspicious, watchful, uncertain as to whether she was a person or a chimp—yet she blossomed into a wonderful auntie to a lot of our younger chimps, and a matriarch of sorts. She is the oldest chimpanzee at Chimfunshi and presides over the sanctuary with a rather royal air. For seventeen years, Milla never saw another chimpanzee; she spent many of those years in a bar, and I can't imagine what went through her mind, but I know it must have taken a very strong will to survive all that.

We first heard of Ludmilla (her full name) in 1989 through Jane Goodall, who informed us that there was a chimpanzee living at a bar in Tanzania and asked if we could take her. I seem to recall Jane's guessing that this chimp was about eight to ten years old; we were anxious to get a slightly older chimp who might serve as a role model for our mob, and Jane seemed very anxious for us to take Milla, so we said yes.

Milla was discovered in a meat market in Cameroon when she was a very tiny baby, tethered alongside the body of her dead mother, and was bought by a very generous British couple, who brought Milla to Kenya and looked after her as their own child until she was about five years old. Unfortunately, the British couple had to leave the country, and left Milla with caretakers in Arusha, Tanzania, where she wound up as a barroom attraction at an Arusha hotel. For years she lived at that bar, becoming addicted to cigarettes and alcohol in the process. That's where Jane came across Milla, and became determined to do something about her condition. Jane apparently considered attempting to reintroduce Milla to the wild chimpanzees she'd spent so many years studying at the Gombe Stream national park in Tanzania, but there were fears that after so many years away from her own kind, Milla might never fit in. So Chimfunshi became the next best option.

Milla arrived in a single-engine plane from Arusha to Chingola, along with Jane and Dr. Ken Pack, a veterinarian from the U.K. Although I was terribly excited and more than a bit nervous about meeting Jane—whose books and films, after all, were literally all we had to go on, especially in the early years—all that was forgotten when I peeked into the box and saw Milla. I thought there must have been a mistake. I expected to see an eight-year-old female; this chimp was balding on top, with sparse, gray hair, yellow teeth, and hands that were gnarled by age. She was also grossly overweight.

I was shocked and blurted out, "Goodness, Jane, she looks older than eight years old."

"Oh," Jane answered softly. "Do you think so?"

Having heard that Milla enjoyed tea, I brought along a jug of it. But Milla refused to touch it, and seemed determined only to rip apart her crate, for not only was she bigger than I expected, but stronger, too. Rather than waste any time, we loaded Milla's crate into the back of the truck, and both Jane and

Ken insisted on riding back there with her, seated on a pile of sugarcane, even though it was frightfully cold. They tried to keep Milla warm by wrapping blankets around her cage, but she went them one better: She pulled the blankets through the bars and simply wrapped herself in them.

It was dark by the time we reached Chimfunshi, and everybody struggled to figure out the best way to unload Milla and her crate, gesturing this way and that with flashlights. The whole scene was rather eerie and a bit chaotic, but Dave managed to organize our guys, and they slid poles underneath the cage and prepared to lift the crate out that way, one man stationed at each corner. But after a single hoist, Dave knew something was up.

"It says in Jane's book," he muttered, huffing and puffing, "that an adult male chimpanzee can weigh as much as a hundred ten pounds." Dave continued to breathe heavily. "But I appear," he said, "to be lifting a subadult eight-year-old *female* chimpanzee that weighs at least a hundred and *seventy* pounds!"

Jane smiled hopefully. "Don't forget, Dave," she said, "Milla's had an awful lot of beer and Cokes."

Later, as we sat out on the verandah enjoying a drink after getting Milla settled in a cage by herself, we all laughed at Jane's "miscalculation" of her age. Jane had wanted so desperately for Milla to get a new life that she was willing to bend the truth a bit, fearful that we might say no, had we known the truth. And we might have—but that would certainly have been our misfortune indeed.

We purposely took Milla's reintroduction to the other chimpanzees slowly, and kept her out of sight of the others at the start. We feared she might be a bit temperamental, especially after being confined for so long, but she really just seemed to be homesick and depressed. Jane spent the next day offering to groom Milla or fetch her a Coke, but Milla wanted none of it. Then, on the second day, she began to get angry, picking and pulling at the wire of her cage, then chipping away the concrete with her fingers. Clearly, Milla wanted out. Dave fetched some materials to begin reinforcing the cage, but before he could, Milla grabbed hold of the metal door at the top, swung her body back, and slammed the door hard with both feet. Twice. The second time, the padlock shattered, and suddenly Milla was free.

We all scurried into action. Ken even got his blowpipe ready with another dart, in case he needed to anesthetize her again, and Philippe Bussi got our dart gun as well, but Milla simply wanted to have a look around. And indeed, she investigated every corner of Chimfunshi. She found a bag of dry cement, opened it, and scooped handfuls of the mix onto a table. She selected two raw eggs from the chicken house and ate them slowly, then wandered over to the verandah, opened the refrigerator, and removed three Cokes. She placed them on a table, hopped up onto a chair, and used her teeth to open the first bottle, drinking the soda with relish. Milla also found a bottle of lime cordial and sipped it thoughtfully, then poured the rest into some dry cement and used a trowel to mix it up. On and on this soft destruction went, with Jane, Ken, Patrick, and Philippe following her about, even leading them on an impromptu bush walk around midday, where they all relaxed on a large termite mound.

It was only when Milla decided to approach the seven-acre enclosure and see what lay on the other side of The Wall that we grew concerned. Although she'd shown no interest whatsoever in the baby chimps we had living in cages near the house,

Milla suddenly perked up when she heard the adults in The Wall begin calling, and she set off purposefully to investigate. Despite her bulk, she hopped right up on top of the wall—a leap of at least six or seven feet from the outside—gathered herself, then began calling to the chimps down below.

Not surprisingly, our chimps reacted angrily at the sight of a newcomer. Several puffed themselves up to enormous size in order to be as threatening as possible, while others grabbed sticks and menacingly began beating the inside of the wall. Milla was surely confused by all this—her first look at a chimp in almost two decades and *this* was the response?—but now our concern also grew for her safety. Ken had shot a dart into her just as she started up the wall, and we feared she might become drowsy, fall in, and be unable to defend herself against the other chimps. Luckily, Patrick was able to scramble along the wall to where Milla was sitting, and he put his arm around her and held her as she nodded off. We used ladders and scaffolds to ease Milla down off The Wall, hoisted her down into a wheelbarrow, laid her head on a sack pillow, and carefully returned her to a reinforced cage.

It was clear from that first day that Milla was like no other chimp at Chimfunshi, and she quickly established herself as a star. Each morning at seven o'clock, Dave would take tea and a bread roll to Milla, and she greeted him in what we came to realize was her own unique style: She would tap her fingers on her chest and make a sound like a low laugh. As we watched her settle into the routine at Chimfunshi, Dave came to christen her the Old Bag—lovingly, of course. With Milla came a battered old stainless-steel jug that she'd had for the last twelve years, which she had used to ask for tea or drinks in the bar. A day or two after Milla arrived, Peter Chilondo, a young man who worked in our garden, came to the kitchen door holding the cup in his two hands and wearing a look of utter disbelief on his face. He had been standing and talking to Milla through the wire of her cage when she handed him the cup, got a grip on his shoulder, turned him around, and, her right arm fully extended, pointed with her index finger at the kitchen and shoved him toward the house. I duly made her a mug of tea, which Peter took back to her.

Milla soon had the entire staff wrapped around her little finger, and it seemed the kettle was always on. It got so bad that I began to have to ration food and treats, lest Milla get them all. Whenever she wanted something, she beckoned the bricklayers or one of the laborers to do her bidding, and they'd go running off as if on an errand. She always got what she wanted, even if it meant that someone's work had to be interrupted. And if Dave complained about a job's progress, the response was always the same: "Ah, but Milla wanted something."

As a result of the attack that left Sandy so outright terrified of the other chimps, we decided that he might be the best candidate to be the first chimp to really "meet" Milla. So, almost two months after she arrived, we got a second cage ready next to hers with a small cage in between and moved Sandy from his solitary enclosure to a new home alongside Milla. Either chimp could enter this small middle cage when its sliding doors were open, then reach through the bars into the other's cage from a ledge, but it also afforded a bit of distance, if necessary. We tried Sandy in the middle initially, but the first time he went close to Milla she put out both hands and shooed him off the ledge. So we locked him back in his side and let Milla into the center area. She took up residence there—moving all her blankets and bedding through the sliding door, obviously deciding that it was a nice place to sleep. She actually made a bed on the shelf alongside the bars of Sandy's cage. We watched intently, but at first, neither seemed terribly impressed with the other. Though Sandy went to the bars when Milla was elsewhere, he seemed very nervous whenever she came close.

Milla, meanwhile, wanted nothing to do with Sandy, and for three days virtually ignored him. But on the fourth day, as I was attempting to give Milla some bananas, Sandy approached to watch. Milla turned to Sandy and put out her right arm, then flicked her hand at him in a gesture that I thought meant, "Go away!" Her hand, when she flicked it, touched Sandy's shoulder, and he jumped away with a little cry. As they looked at each other intently, Sandy began to move, very, very slowly, closer to Milla. She again put out her arm and flicked her hand. This time, Sandy did not jump away when it touched him on his shoulder. Milla put out her arm a third time, flicked her hand, touched Sandy's shoulder, and this time gently took hold of a handful of his fur. Sandy seemed frozen to the spot as Milla moved her hand, in a very slow motion, up and down his left arm. Seventeen years after last touching another chimp—probably her dead mother—Milla was stroking one of her own kind.

This seemed to go on forever, and then Milla moved her hand onto the top of Sandy's head and very gently pulled him closer to her. There seemed to be no reaction, but I could feel the atmosphere intensify. Only when she put forward her pouted lips and kissed him on the top of his head did he seem to relax. Suddenly, Milla put out her arms and pulled Sandy toward the bars, at the same time pushing herself up as close as possible so that their bodies touched for a brief hug. She then started to examine Sandy all over—his head first, then his arms and hands and on down to his feet. She tried to touch his penis, but he closed himself up and would not let her—perhaps he was still tender from the attack. But Milla was satisfied and began grooming Sandy's neck instead.

I was watching so intently that I did not realize at first that there was very quiet laughter coming from both of them. Then I noticed that the rolls of fat on Milla's back were wobbling in rhythm, and that Sandy was chucking softly, too. Sandy slowly put his hand out to touch Milla's cup, but Milla picked it up and moved it out of his reach. He then tried to touch one of her bananas; she moved them out of his reach. And it was the same with her orange. So he picked up one of his oranges and gave it to Milla, and she accepted the gift, thereby forging an unmistakable bond.

A few weeks later, we opened the inner doors between Sandy and Milla's cages and the two of them began to move freely back and forth. But there was not much interaction at first. Milla walked right past Sandy and went into his cage to get a better look at what Dave was doing off in the distance, while Sandy strolled right past Milla and took the liberty of examining all her belongings—her cup, her blankets, a doll, a ball, and a pink plastic toy, the last of which he started playing with. They seemed oblivious to one another for the first thirty minutes, and then, suddenly, they were together in one cage, grooming, examining, and eventually having an afternoon's rest close together, as though they were the best of friends. Milla and Sandy seemed to satisfy something in each other—he a rowdy, wayward adolescent, she a lonely old maid—and their relationship, especially in the early days, was beautiful to watch. Where Sandy was all mischief and pluck, Milla was quiet and dignified. Often, he'd spend hours looking for things to throw at the geese or peacocks

(Milla's cup was a particular favorite), and he loved walking backward and forward and turning somersaults right in front of Milla, as if trying to annoy her. She, of course, ignored him. They even shared food. Sometimes Milla would take something from Sandy that she was rather partial to, like a cucumber or green peppers, and Sandy would make no protest. And when we gave them their tea in the mornings, Sandy would often take a sip from Milla's cup as I passed it to her, and she would just gently move his head out of the way.

When there were disputes, however, Milla was quick to put Sandy in his place. Once, when I went to feed them, I found Milla was chasing Sandy round and round their cages, screaming at him. She caught hold of him and appeared to bite him on the back, then threw him onto the ground and gave him a very vicious shaking. Milla then climbed onto the table, turned to me, and held out her hand for her food. Sandy stayed on the floor for a few seconds, crying and whimpering, and then —still crying—crawled slowly up onto the table next to Milla and leaned toward her. Instinctively, Milla reached out to him and, in a few hasty gestures, brushed the grass off his back and patted him on the shoulder. Sandy immediately stopped crying and picked up a guava to eat.

Because she spent so many years listening to humans and responding to their voices, I'm convinced Milla understood Dave and me like no other chimp I've ever met, and there were times when she even seemed to "side" with us humans. When Sandy once stole a pair of pliers that Dave had accidentally put too close to the cage, Dave held out his hand and said three times, in a quiet voice, "Sandy, please give me my pliers." But Sandy kept dancing out of reach with the pliers in his hands, clearly enjoying this new game, as Milla looked on.

Then Dave said in a loud voice, "I want those pliers now." Still, Sandy danced around with the pliers.

Finally, Dave shouted, "Hey, you! Come here with those pliers!"

At that, Milla ran after Sandy, caught him, snatched the pliers out of his hands, then gave him a slap across his head. Milla handed the pliers back to Dave, and I do not know who was more surprised—Dave or Sandy.

Introducing Milla to Sandy constituted only half of her rehabilitation, however. For many reasons, neither she nor Sandy could be placed into the family group of chimps living in The Wall, so we created a second, larger enclosure, one encompassing fourteen acres that was built directly in front of the house, just a hundred yards from our verandah but on the opposite side of the main access road from The Wall. Like The Wall, it had a maze of interlocking handling cages that allowed different chimps access to the enclosure at different times, but the big difference was that the fourteen-acre enclosure would be ringed by a ten-foot electrical wire fence. Given the woeful Zambian economy, bricks and mortar were far too expensive to ever use for chimp enclosures again, but the advent of reliable solar-powered generators allowed us to use electrical fencing at a fraction of what it would have otherwise cost. It also meant that the chimps could see everything through the wire, including the chimps up in the trees behind The Wall.

Naturally, the addition of a new enclosure forced us to hire more chimp keepers, and we now had a staff of about a dozen Zambians who handled everything from preparing food and cleaning cages to taking the youngest chimps on their daily bush walks and helping Dave build new cages. We were also fortunate over the years to have our five children return to help

My brother Keith and me in North Africa, midway through our trip

My brother Clive with our dog, Tim, in the Sahara Desert in 1947

This photo was taken in my early twenties.

I loved rally car driving, even though women were frowned upon in the sport. Here I am in 1953.

Dave with Pal, shortly after his arrival in 1983

Pal had the run of the house, and often took sips of our beer.

Boo Boo, the troublemaker

Donna adored Boo Boo but hated humans.

Chiquito's incredible strength was evident right from the start.

Philippe Bussi operated on Sandy (with me assisting) every day for weeks after the attack; there's no doubt he saved Sandy's life.

Sandy's bruised and battered face after surgery

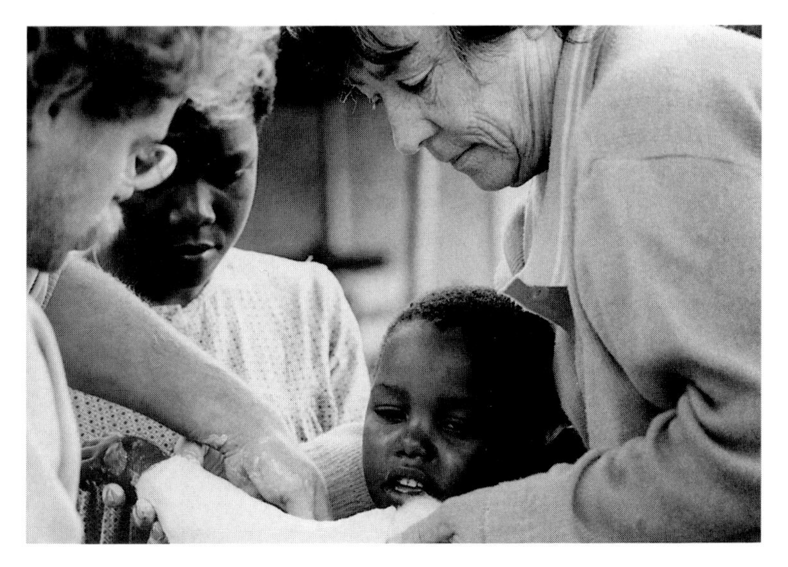

When word got out Chimfunshi had medicine, we became a virtual field hospital. Here, Philippe and I mend a child's broken arm.

Patrick Chambatu and Jane Goodall became good friends.

Noel (left) grooms Choco, the sort of typical chimpanzee social behavior that Chimfunshi tries to foster.

Dave comforts Stephan, a young male who arrived from Russia in 1997.

The youngest orphaned chimps are taken out each day for bush walks with keepers.

Billy the hippo preferred people to her own kind and celebrated Mass on one memorable occasion.

The collapse of The Great Wall led to one of the most chaotic chimp escapes we ever endured. Here, my sons and I survey the damage.

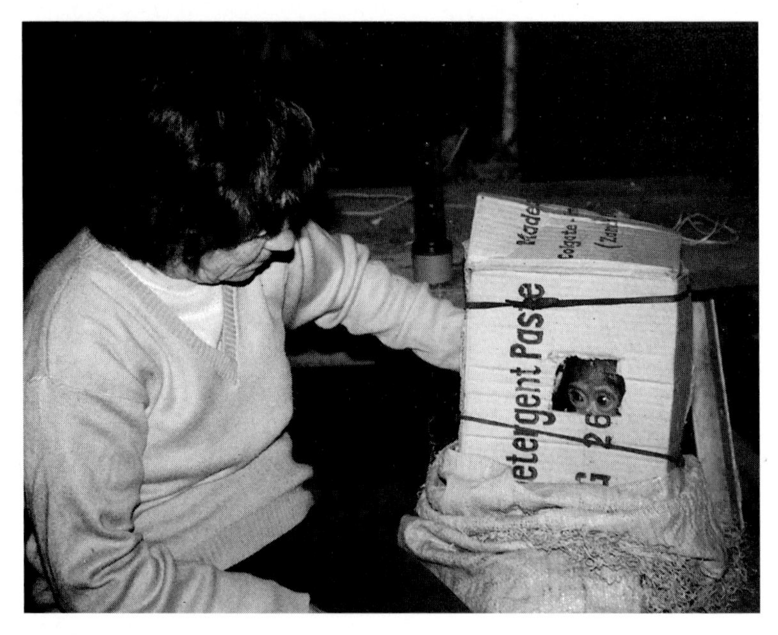

Chimps came to us in a variety of cages and containers. Pippa arrived one night in a cardboard box, with only her eyes peeping out.

My beloved Sheena, whose trust briefly overcame her injuries

Cradling Doc shortly after he was confiscated from smugglers

out at different times, often living for years on nearby farms or in Chingola, and each of them was invaluable. Indeed, it was just as Chimfunshi's growth was threatening to truly engulf us that we were lucky enough to have our eldest daughter, Lorraine, return with her husband, Ian Forbes, and their three sons. They settled in a farmhouse we'd built about two miles away, around a bend in the Kafue River, and became vital members of the sanctuary. While Ian took on the role of ranch foreman and created a campsite, Lorraine became my primary helper, and she quickly developed into a chimp expert.

When we decided to release Milla and Sandy into the fourteen-acre enclosure, we thought it best to let them roam the property alone first. As Dave and I waited inside the new enclosure in order to coax them out, Sandy appeared through the opening and rushed up to me for a good, long hug. Milla, however, put her head out the door and slowly looked around before stepping gingerly through the opening. I must say, I was struck by her appearance: Milla had lost a great deal of weight since arriving at Chimfunshi, and her old rolls of fat were now muscle that contoured properly to her ribs. But because she had not spent much time in the sun, the creases between her rolls remained very white, and she looked like a skeleton from the back.

Both Milla and Sandy had spent a few days in the handling cages before we released them, and it was clear she had looked long and hard at the path leading to the large acacia trees in the distance. She grabbed my hand and, as she made happy grunting noises to herself, we headed toward the trees. But the chimps in The Wall could see Milla from their trees off in the distance, and a chorus of angry warning sounds went up. Milla changed direction and started toward the noise, and before we could stop her, she'd gotten to the electric fence, put her arm through, and grabbed the earth wire on the other side. She touched her mouth to the electric wire, and suddenly Milla's body jumped off the ground a little as the voltage hit home and she began screaming. She bolted from the fence, shrieking, and seemed to move in different directions, not knowing what had hurt her. She turned to Dave, and it was awful watching her stagger toward him, still screaming for help, reassurance, any kind of comfort. Dave hugged her and spoke softly to her, and after quite a while, Milla calmed down.

Despite that rude start, the next few hours were wonderful. Milla climbed a large anthill and surveyed her new world. She did not sit but just slowly moved this way and that, all the time looking. She wanted to walk, so we walked. She wanted to sit, so we sat. It was as if she could not decide what to do first; her sense of excitement and wonder was palpable. Sandy was behaving as only Sandy could, either making himself obnoxious or begging me to carry him. After about an hour, we were again sitting on the anthill, with Sandy scrambling overhead in the trees and teasing Milla. She suddenly jumped up when Sandy passed by and tried to grab his foot. Then, in her slow, clumsy way, she started chasing Sandy around the top of the anthill and in and out of the branches of the tree. They were both laughing and had play-face expressions, and I stood there for at least ten minutes taking it all in. I think Milla stopped only because she was exhausted, tired from using muscles that she hadn't in years.

Unfortunately, Milla and Sandy's happy days together in the new enclosure were short-lived—and Sandy was the reason. Two days after we released them, Dave and I were standing around when I heard him say, "Oh no." I looked toward the enclosure and there was Sandy, walking across the compound in our direction. Nobody saw him get out, and there were no clues as to where or how he'd accomplished his feat, but it was the first of many, many, many escapes over the next decade that have made Sandy the most frustrating chimp at Chimfunshi. We locked him in a handling cage for the next three days and tried to find his escape route, but to no avail. When we released Sandy again, he quickly greeted Milla, gave her a big hug, then turned and escaped once again. The whole episode had taken no more than five minutes, but one of our keepers, Wilson Lupanda, saw it all clearly. He said Sandy had used the bars of the cages to swing himself up high enough to get a two-finger grip at the edge of the wall—a leap of eleven feet. Amazingly, those two fingers were enough for Sandy to hoist his entire ninety-pound frame up and over, and once again we were at a loss as to what to do.

Clearly, Sandy could not be trusted in the new enclosure—yet Milla could not be left in there alone. Luckily, an influx of newcomers to Chimfunshi suddenly filled most of the adjacent handling cages, allowing Milla to integrate slowly with a whole new community. In one cage we had Toby and Lucy, a pair of chimps rescued from a bankrupt circus in the South Pacific, and Choco and Leben, half-brothers who joined us from the overcrowded Tel Aviv Zoo. Four recent arrivals from Rwanda—Misha, George, Georgette, and Maggie—occupied another cage, and the six youngest chimps, whom we took to calling The Babies, were in yet another cage. When Milla came in for lunch one day, I remember thinking how amazing it was to see her sitting and eating, turning her head now and then to look at chimps everywhere around her. Even though there were bars separating them, the chimps could all see one another, and for Milla it must have been the first time she'd eaten with her own kind for nearly nineteen years. It was clear, too, that she had been craving the company.

Milla appeared to be most taken with The Babies in the cage next to her. Most were about two years of age and, in addition to the Rwandan chimps, all arrived within an eighteen-month period that was one of our most active at Chimfunshi. Mike and Grumps were a pair of males who were found in a small wooden box drifting around on the baggage claim conveyor belt at the Nairobi Airport, while Pan and Dora arrived after a horrifying journey that took them from Uganda to the Middle East and back again. Goblin had been stuffed inside a tiny crate for so long that his fur had turned red and his legs were nearly paralyzed, and Pippa was actually brought to us by poachers who offered to sell her for $2,000. We told them to wait so that we might go to the bank and get some money, but we returned instead with the police, who arrested the poachers on the spot. Pippa, meanwhile, took in the whole ordeal from a rumpled cardboard box, and all you could see through a tiny slit were a pair of huge, terrified eyes.

Milla would sometimes lie on the ledge next to their cage and pretend to be asleep, allowing the braver souls like Pan to sneak up and grab or bite any part of her that seemed to stick through the bars—a finger or foot, or even just a toe. Though Milla was fully awake, she never showed any reaction to this nasty behavior. In fact, Milla often encouraged Dora to approach and groom her through the bars, sometimes even turning her back, and with a finger indicating to Dora a spot she'd like scratched.

About this time, we received another older female chimp named Noel, who arrived as a political refugee of sorts from Zaire. There had been a lot of fighting and looting there due to political unrest, and expatriates were told they should evacuate. Noel's owners, who had kept her as a pet for the past twelve years, knew they had to leave and could not continue to care for a chimpanzee, so they sent her to us. Noel is a lovely chimp, and we believe she must have been about fourteen years old when she arrived. She didn't seem to be the least bit aggressive, and wanted only to receive and give back love. We put her in a cage alongside The Babies for a couple of weeks in order to get her acclimated, and since they all seemed quite fond of one another, I eventually let them in with her. Noel was wonderful. Everything she did was done slowly and carefully. The Babies were a little nervous around her at first, but Noel did not rush at them or go near them if they seem frightened. She just sat in the cage with them and let them come to her.

After Noel was clearly established as The Babies' protector, I opened the doors one day and let them all into the enclosure with Milla. Though Milla seemed confused and approached them very cautiously, one of The Babies began barking, and Noel responded by standing up to her full height and puffing out her hair to make herself look as big as possible. Milla made a threatening move, but backed off when Noel began walking toward her. Milla beat a retreat and started to climb a tree, and Noel decided that this was a good time to start the chase, and followed close behind. They scampered up a large acacia tree, then sat glaring at each other with their fur standing on end, but with neither of them making any noise. Noel and Milla must have stayed in the tree for about fifteen minutes, but then seemed to get restless at the same time. Noel started down first, and with every foot that she moved, Milla also moved a foot. At last they were on the ground, keeping a wary eye on each other, and Noel slowly began backing toward the cages. Milla picked up a thin stick that was about four feet long and had a few dead leaves at the end, and waved it above her head while barking at Noel. Milla brandished her stick this way and that, but Noel was satisfied that Milla posed no real threat and nonchalantly returned to the cages.

After that, Milla blossomed in the fourteen-acre enclosure, and began to display a number of curious habits. For instance, she is one of the few chimps I've ever known who seems to prefer walking upright on two legs, rather than on her knuckles, as most primates do. Even when she walks alongside other chimps, Milla often toddles back and forth upright, whether her hands are full or not, and one wonders if this might not be a trait she acquired from humans. Sometimes, Milla strolls about picking long stems of grass, and to me she looks just like a woman in her garden picking flowers. In the mornings when we give out the maize balls, Milla comes with her small bunch of grass to collect her portion. With the maize balls in one hand and the grass in the other, she makes herself comfortable and takes a bite of each in turn.

Milla also carried a blanket in the early days, a tattered old piece of cloth that went everywhere she did. Each morning she would pick the blanket up and give it a good shake, then spread it out neatly and sit down upon it rather than get her bottom wet from the dew on the grass. At night, when most of our chimps headed into the trees to make nests, Milla would invariably get into her box, put the blanket smoothly under her, then lie down on it and pull a burlap sack up over her shoulders. But this blanket had other uses as well. Once, I watched her flick the blanket through the bars of the cage and hit a dog that had his nose in the chimps' food basket, and other times we saw her flick it at geese or peacocks if they were out of reach. But our biggest surprise came the time we watched her put six small sweet potatoes, two guavas, and an orange in the middle of the blanket, very neatly fold the corners into the middle until she had a small parcel, and then pick it up and carry it off to a comfortable place to eat, away from the other chimps.

When Jane Goodall returned to Chimfunshi in 1995 to visit Milla, a blanket again played an interesting role. We wondered if Milla would recognize Jane or not. At first I thought she did not, because the chimp looked hard at her and then turned her head away. I think Jane was a bit disappointed, but then I noticed that Milla was looking at Jane almost out of the corner of her eye, and seemed to keep staring at her in that way for a very long time. Suddenly, Milla turned and gave Jane her usual greeting—tapping her chest repeatedly with her right hand and making that hoarse laugh—as if she'd just placed the name with the face.

Jane asked if she could give Milla a blanket, as Milla had recently lost hers. I sorted out two sacks—a slick nylon and a burlap, like the one Milla had lost. We went to the fence, and Jane called Milla over, talked to her for a while, and threw the nylon sack over the electric wire. Milla looked at the sack for a moment, then turned her back on it. Jane again spoke to Milla, but this time threw the burlap sack over the wire. Milla immediately picked it up, walked a few paces, shook it, and then smoothed it onto the ground and sat on it. As Jane left, Milla watched her intently until she was out of sight.

At feeding time one day, George, a headstrong young male, came toward the handling cages waving a five-foot metal drill rod above his head. I assume he'd found it after a careless worker left it behind, but I was very worried that he'd hit someone on the head with it or catch it in the electric wires of the fence and electrocute himself. He was really showing off—rushing here and there with the rod. He eventually got a bit tired of this game and threw the rod with great force into the grass. We could not go in and retrieve it just then, so Lorraine and I thought we would try and get Milla to get it for us. But she had not seen George with the rod, so she did not understand what we were asking her to get. We offered her a box of chocolates and some apples and indicated for her to go toward the long grass. She slowly walked toward the grass on two feet, but kept looking back at us, almost asking if she was doing the right thing. After looking for some time, Milla got fed up and started to return to the cages. Lorraine and I waved her back, shouting, "No! No! Milla—that way! That way!" She eventually spotted the rod in the grass, picked it up, and carried it to the cages, where she pushed it through the barred window. Then she held out her hand for her reward.

We never did find out where the rod came from, but two days later, Milla brought us another metal drill rod she'd found in the enclosure. As before, she shoved it through the bars, then held out her hand for a reward.

It's also clear that Milla viewed humans as her special protectors whenever she got into a dispute with other chimps, but as I saw when we had a BBC television crew here to film a documentary, she was also ours. The crew was lovely, all animal people, and their visit went very well until the end, when they wanted some interaction with a couple of the bigger chimps. I thought Milla and Chiquito might be a good bet, as they seemed to be alone in the fourteen-acre enclosure at the time—but I was wrong. Milla was full of love and greetings when I entered the enclosure, but suddenly out of the corner of my eye, I saw Pan running toward me. This upset Chiquito, who lunged forward and flipped me up in the air, knocking me off my feet. I fell awkwardly, breaking my wrist with a loud *snap!*

I realized I was hurt and struggled to come to my senses. When I did, I was shocked by Milla's reaction. She was *enraged,* and immediately began screaming at Chiquito and running after him, at first on two legs, then all four, her yellow teeth flashing and her thin gray hair bristling. Milla chased Chiquito a long way into the bushes, all the time screaming at him with a horrible, piercing cry. I set the wrist myself and was back at work the next day, but Milla would not leave Chiquito in peace. For the next three days, she screamed and flicked her arms at him whenever he came to the cages for food, as if telling him that he was not welcome, and Chiquito genuinely appeared to be frightened. In fact, he kept well out of her way, running off into the trees every time she looked his way.

Ultimately, Milla was forced to fight to establish her place within the social group. Like all chimpanzee fights, it seemed to explode out of nowhere. Milla acted first, attacking Cleo, a young female who'd proved quarrelsome since she arrived in 1986, over a guava that Milla probably thought was hers. Milla jumped at Cleo and tried to bite her on the shoulder, but Cleo started hitting Milla on the head. Milla got furious and started her high-pitched screaming, at which stage everyone else joined in. Milla rushed to the bars of the cage and asked Dave and me for assistance, while Cleo went up to Chiquito and asked for his. Chiquito just sat with his arms folded, figuring the odds. Dave and I would not help, either, of course, so the two girls got pitched in again. We could not make out who was who as the two of them rolled furiously around on the floor. The other chimps were all on different shelves, jumping up and down, making terrible noises as if offering encouragement.

Milla and Cleo finally split, but then Cleo picked up a cornstalk and started beating Milla over the head with it. Milla picked up a corncob and threw it at Cleo, then they both charged again and tried to tear each other apart. They split once more, but Cleo found another maize stalk and started hitting Milla, who was grabbing armfuls of straw and throwing it at Cleo. The fight went on for a good fifteen minutes, at which point the two of them were so exhausted and breathing so heavily and quickly that they could not carry on. They practically crawled to different sides of the cage and glowered at one another, still screaming. Finally, Chiquito blew himself up to his full, impressive size and rushed at Milla, who seemed to cower. But he just danced around her a few times, made it clear the fight was over, then went back to his shelf and sat down. The other chimps kept adding their very shrill screams long afterward—I cannot remember the noise ever being worse. My ears were ringing, and I had a headache for the rest of the day.

In many ways, though, that fight hastened Milla's assimilation into the Chimfunshi chimpanzee community. After months of being sort of standoffish and bad-tempered toward the other adults in her group, Milla suddenly seemed to decide that she had better try and make friends, in case she needed an ally in times of trouble. For the first time ever, she started to make advances to Toby and Leben, two teenage males who she thought might one day challenge for dominance. Standing to her full height, Milla put on a play-face one day and began grooming Toby as she chuckled to herself, then half-turned to Leben and patted her chest in her special greeting. I think the two guys were a bit overwhelmed—they just sat and looked at her, stupefied—but she was very persistent and started grooming Toby's feet. I think he liked it, eventually.

Milla even began to show interest in sex, an act that must have been completely foreign to her. Female chimpanzees reach sexual maturity at about eight years of age, and often give birth to two or three offspring by the time they are eighteen years old. But at that age, Milla hadn't seen another chimp—let alone ever had sex with one—in so long, she'd probably forgotten what they were, so her amorous attempts would have me roaring with laughter. One day, Chiquito was lying on his back, resting in the sun, nonchalantly watching the others play, when Milla walked past him on two legs, then stopped and took a long, hard look at him. She promptly went into reverse, sat on his lap, and patted her bottom very invitingly. This position lasted only about thirty seconds, but since Chiquito took absolutely no notice of her—if anything, he was in shock—she got up and continued on her way. Milla regularly comes into oestrus like the other females, but the male chimps refuse to mate with her, perhaps out of respect, perhaps out of bewilderment. Experts say that a female chimp can reproduce until very late in life, but I doubt that Milla will ever be one of them.

Nevertheless, even though she's not living in the jungles of Cameroon with her proper family, Milla today is safe and happy and well adjusted, living with a family of chimps who need her just as much as she does them. Recently, I saw the funniest thing—Milla with her hands full of bread and rolls standing on one leg, leaning against a wall, while using her other foot to tickle the tummy of one of the infants, Dolly. Dolly, who also had a bread roll in one hand, had hold of Milla's foot with the other and was laughing so hard that she was gasping for breath, and Milla was laughing, too. Soon, the other chimps stopped to watch, and began laughing along. Everyone was laughing so much that none of them could stop to eat their bread. The sound was wonderful.

Nine

Escape

Charley, Chimfunshi's first dominant male and the leader of so many escapes

From the moment we began housing our chimps in cages, one fact became abundantly clear: "Escape-proof" did not exist. In fact, it was never a question *if* our chimps would escape from this cage or that enclosure, but *when* the breakout would occur. Even though we'd come to hold tremendous respect for their intelligence and creativity—let alone their incredible strength—it was the chimps' patience that left us in shock. They'd sometimes spend weeks picking at a weak spot in a fence or a soft patch in a wall, or simply twist an insignificant piece of wiring on a cage day after day until at last a hole opened up big enough to squeeze through. Some experts told us that chimpanzees only seek to escape their surroundings if they are unhappy, but I doubt this is true. Our chimps seemed to take delight in escaping simply to see if they could. Sometimes, they would even let themselves back in through the very hole they'd used to get out.

Liza Do Little was our first real escape artist at Chimfunshi. When she and Charley became too large and unpredictable to be trusted out on the daily bush walks, they were placed in a large cage together until the seven-acre walled enclosure could be built. Although Charley was delighted to have his best girl so close at hand, Liza spent all of her time looking for ways to get out. Early one morning, she succeeded, ripping apart a small section of wire mesh that was just big enough for her to wriggle through, and off she went into the nearby fruit orchards. But the hole was too small for Charley, and when he realized that Liza was gone, he grew so angry that he began smashing the planks of wood that formed their sleeping platforms, crying the whole time. I went and sat with him and held his hand, and I even spent a large portion of the day grooming him through the wire of the cage. This seemed to calm him down a bit, but when Liza reappeared around three o'clock in the afternoon, just sauntering lazily through our compound as if nothing were amiss, Charley went crazy again. We opened up the cage door and she strolled in, only to be engulfed by Charley, who threw his arms around her and refused to let go.

For the next hour or so, Charley refused to take his hands off Liza. And when she stepped up into the sleeping area, Charley was close on her heels. They usually made two separate beds close together, but on this day, Charley would have none of that. He got into his bed, grabbed Liza round the neck, and pulled her close. He put both his arms around her and kissed her three times on the top of the head, and would not allow her to budge for about twenty minutes, then finally let her get up and go for a drink of water. But Charley kept motioning for her to return, and did not rest until she did.

Not long after, Liza escaped again. We saw her go off to the orchard and thought she'd come back when she was ready. We were fine with her being out all night, even though it meant Charley would not sleep well. But just after sundown, it started raining very heavily, and the night was very dark.

At about seven o'clock, Dave and I were reading in the lounge when the front door burst open and Rita, Tara, and Donna rushed in, closely followed by Sandy, Little Jane, and Coco. Liza had apparently made a hole in The Youngsters' cage while attempting to let herself back into her own cage, and now we had six soaking-wet chimps racing around the house, bouncing off the furniture and clattering down the hall. It was a real battle to get them back into their cage, especially as the intermittent claps of thunder seemed to occur only at crucial moments in the transfer, sending all six racing back toward the house. Dave was soaked to the skin while repairing the holes in the cages in the pouring rain, with only a flashlight to work by, and Liza was nowhere to be seen. We finally sorted everything out satisfactorily and went to bed about ten o'clock, by which time the storm was really bad, with lots of forked lightning and frightening explosions of thunder. Nevertheless, we somehow drifted off to sleep.

Around 10:45, we heard what sounded like someone trying to batter down the front door. We both shot out of bed and rushed to the front of the house, realizing that it must be Liza trying to break in. I opened the door and called out into the rain to her, but she would not come to me. Then I heard the back door being pounded on and rushed to it, but as soon as I opened it she sped away again. I never actually laid eyes on Liza that night, but it was clear she was close at hand, alternately thumping on the front and back doors for the next five hours. Neither Dave nor I got any sleep, and the storm didn't peter out until at around five o'clock in the morning. As soon as it was light, I went out and opened the door to Liza's cage. She materialized out of nowhere and shot past me into the cage, where she was received joyously by Charley.

When we moved the chimps into the seven-acre walled enclosure, we hoped the added space and lack of human intrusion would make them forget all about escaping. But if anything, The Wall became their new enemy, and finding ways over, around, or under it seemed to be the only thing on some of their minds. Liza, of course, was at the top of that list. The chimps had not been inside the enclosure for three months before Liza had found the weakest spot—down by the riverbank—and began making regular excursions out into the compound, often accompanied by either Charley or Spencer. The walls of the enclosure extended right down into the Kafue River, where one whole border of the enclosure was a two-hundred-yard stretch of river. During the rainy season, we could expect the water level to rise to about twelve feet, and since chimps cannot swim, we knew that they would not try to get around the wall. The trouble began when the heaviest rains failed to come and the river rose to only about six feet, leaving Liza plenty of room to walk out past the end of the wall to freedom.

Dave spent several days with some of the farmworkers chopping down branches from the thorny acacia trees on our property and piling them out past the end of the wall. This would have stopped most chimps, but not Liza. With infinite patience and precision, she spent hours among these piles, carefully breaking away just enough branches and thorns to fashion a tunnel under the brush. Liza would then wait until no one was looking, stroll out to her escape hatch, and wriggle through to freedom.

An escaped chimp is dangerous and unpredictable, and our less-experienced staff usually disappeared from sight the minute Liza or Charley or Spencer appeared in the compound. As a result, Dave and I were usually left to corral and coax the chimps back to their cages alone, a job that, even with Patrick's help, eventually took up more and more of our time. One day, Liza came around the wall with Charley, and after raiding the orange orchard and wrestling with the refrigerator on the front porch, they returned to the enclosure, climbed the outside wall, and dropped back inside. Charley immediately went to the cages, looking for something to nibble on, but Liza wasn't finished. The next time we looked up, she was ambling up from the riverbank with Spencer in tow, then veering off through the cattle paddock and into some very thick forest. They spent the night out before reappearing the next morning at seven o'clock for their milk. In the meantime, Dave and some workers had spent hours putting more acacia thorn trees around the wall, so when Liza returned to the enclosure, she found her escape route sealed off. But within a week, she was out again. All we could do was sit tight and wait for the rains to raise the river.

Charley was always our greatest concern during an escape, of course, but he could also worry us by never leaving the enclosure. We once had a visitor who wanted to paint some pictures at Chimfunshi, and he chose to work atop the viewing platform at The Wall. Unfortunately, one of the chimps escaped that day and stole five tubes of paint, which he carried back into the enclosure and promptly spread everywhere. Two tubes were recovered intact, but I found one bitten in half and another with teeth marks in it; the fifth was never found. Charley also disappeared from sight at this time—missing his meals for four straight days, which was unheard of—and we became very worried. We searched the enclosure on three occasions, and I was terrified he'd either drowned or been taken by a crocodile. But there were no signs of a struggle along the riverbank, and we kept searching.

Finally, on the fourth day, I went into the enclosure in the late afternoon and wandered around, calling Charley's name. I was near the river when I heard faint noises, and found Charley lying against the wall, too weak to move. He looked awful and was begging me for something. I left him and returned with some water and slices of oranges and apples, and he drank the water quickly and devoured the fruit. He motioned for more water, so I fetched it, and he drank at least two gallons and ate more oranges before taking a break.

There were no visible signs of injury on Charley, but it did not take long to put two and two together. Being the dominant male, Charley would surely have demanded the tubes of paint from the younger chimps, and the first thing any chimp does with a new item is taste it. We believe the paint might have contained lead, and given that one whole tube was never found, Charley was probably suffering from a serious case of lead poisoning.

I spent five more days ministering to Charley in the enclosure, bringing him water and food, but I was unable to leave anything, because the other chimps were behaving horribly toward him. The females, notably Liza and Little Jane, stole his food, and Tara decided this was his best chance to dethrone Charley as the dominant male. Tara attacked each of the males, chased and bit many of the females, and even turned on his old friend, Boo Boo, who was so scared that he ran from Tara on sight. Clearly, Tara was seeking to establish a new order before confronting Charley himself. When Charley finally mustered the strength to stagger up to the cages after nine days down by the river, Tara decided it was time for a showdown. He puffed out all his hair, worked himself into an absolute frenzy, then rushed at Charley with a look of determination on his face. Amazingly, the females, led by Liza, Girly, and Big Jane, suddenly banded together to protect Charley and kept Tara at bay, waving their arms frantically and screeching at him in their loudest voices every time he got too close. Tara made run after run at Charley, but never actually laid a hand on him. Eventually, he gave up and went off to sulk, glaring at the females.

It was almost two months before Charley began to regain his strength and resume his duties as dominant male, but when the adult chimps weren't escaping and causing problems, the littlest ones took over, since even the tiniest hole was big enough for them to get through. One day, Dave and I were over at The Wall showing some friends around, and as we walked back to the house, I said I would make us all a cup of tea. But when I opened the door, the kitchen looked like what you see in films when someone's house has been searched. Everything was in a terrible mess—there were buckets strewn on the floor, two bottles were smashed on the table, there seemed to be paper and curtains and empty sacks everywhere, and the floor was covered in half-eaten guavas. I stood there and tried to take it all in, but I couldn't—I was in shock. Just then, some movement caught my eye and I saw a little black furry bottom disappear around the edge of the door to the lounge.

I tiptoed through the kitchen and peered around the edge of the door and burst out laughing. Six of The Babies—Pippa, Mikey, Grumps, Goblin, Pan, and Dora—were all sitting in a clump and hugging each other with looks of absolute fear on their faces. They had got such a shock when I opened the kitchen door that they were all frightened. When they saw me, however, they rushed forward and all cried to be picked up, then began squabbling among themselves as if to assess blame for the raid. We got The Babies back into their cages without any protest, but then the cleanup started. The worst was the guavas—the chimps had discovered a two-gallon bucket full of the fruit, but they were unable to eat it all, so what they had not consumed had been squashed and rubbed into the carpet, as if they'd been attempting to hide the evidence. It took me hours to clean it all up, but at least they had not injured themselves on any of the broken bottles.

On a morning not long after, I was in the kitchen when one of our cleaning women, Sarah Kabinga, ran in screaming, "Charley is out!" She was holding her nearly naked year-old child, Louis, in the air by one arm, and had a look of genuine terror on her face. She had been sitting on the grass changing her baby's diaper when she looked up to see a huge chimp standing next to her, watching. The size of him must have made her think it was Charley, but in fact it was Chiquito. He just looked at Sarah and then the baby, gave a bit of a shrug, and walked away. I calmed Sarah down and locked her and Louis in the kitchen for safety. Then Dave and I went off round the house, locking doors and refrigerators so that Chiquito could do no damage. But as I was fastening the bolt on one of the windows, I looked out and saw Cleo walking down the road. We had no scheduled visitors to worry about and our staff was suitably secure, so Dave and I were very calm.

For months we had been worrying about how we were going to effect the transfer of Chiquito and Cleo from The Wall to the new fourteen-acre enclosure—neither had fit in with Charley's group and needed a new start. Now the two chimps had made up our minds for us. As I went to the new enclosure to make sure that a cage was available, I heard a sound that made my stomach drop: an automobile engine. A car was headed up our driveway with what appeared to be a full load of tourists inside, so I rushed outside waving my arms and got the driver to pull over. I breathlessly explained to the six people inside that we had a chimp or two on the loose and that it might be best if they turned around and waited near the main road, but before they could answer, I looked up to see another car at the gate, with Chiquito hanging on to the open window and a woman with three children screaming inside. I ran to that car and patted Chiquito and told him to go away—which he did, though not before the car had reversed and sped off.

Dave darted Chiquito to anesthetize him, the only way to safely get him back into a cage. We placed him in a wheelbarrow and drove him to the new cages, then laid him out on a bed of straw. Certain that Chiquito would be all right, we set off in search of Cleo. Suddenly, visitors seemed to materialize out of nowhere—I could already count five cars waiting outside the main gate, with at least two more rumbling down the road. But worse than that, I could see people getting out of their cars to set up their picnics: fathers, mothers, and children. My eyes were darting from one family to the next when suddenly I saw a group of children clustered around in a tight circle, with Cleo seated right in the middle. She was loving all the attention, but my heart nearly stopped beating on the spot. I ran up the road, getting angrier and angrier as I got closer, to the extent that Cleo dashed off when she saw me coming. But it was the people I was most angry at, not the chimp. I told the parents that they were being stupid and that they could get themselves bitten for all I cared, but that to expose their children to such a risk was unacceptable.

While I was still giving the tourists a tongue-lashing, I heard a tremendous commotion coming from the fourteen-acre enclosure, so I raced back up the road in that direction. Cleo had jumped on top of the outer wire fence and swung over the electric wire into the enclosure, landing right in the middle of Noel's daily play group with The Babies. By the time I reached the scene, Cleo and Noel were up the same tree, barking hoarsely at one another, while eight of the Babies raced back and forth on the branches between them. As I stepped inside the enclosure, I immediately had three other Babies clinging tightly to my legs, terrified by all the commotion. With no other options, I simply sat down on the grass to watch Cleo and Noel's battle. Cleo kept approaching Noel, who warily kept backing off. Then Noel barked and threatened Cleo. Cleo again approached in what I thought was a friendly gesture—at one point, they actually touched—but one of the Babies barked and bit Cleo on the leg, so she backed off.

Cleo at last came down from the tree and Noel began chasing her, with The Babies in hot pursuit. Cleo ran around the anthill and came straight toward me, perhaps looking for protection, but The Babies on my legs barked in alarm and Cleo changed course. At this point, I thought it best to get out of the enclosure because I was interfering—not to mention in very real danger of getting hurt—so I peeled the chimps away from my body and stepped back outside.

The Babies decided to follow me, so we got them into a cage, then allowed Noel inside and locked the gate. We left Cleo to look around the enclosure by herself for awhile, and one of our workers remained on guard in case she ventured into an open cage. It was about fifteen minutes later that one of our staff came to us and reported that Cleo was now in a cage next to Choco and

Leben. Dave and I went out and commiserated with her, then congratulated each other on a daunting hurdle that had been successfully—if a bit awkwardly—cleared.

Chimfunshi's rapid growth was taking a toll, and we struggled to keep up with the influx of chimps. We averaged about three chimps per year for the first five years—a number that was difficult enough to accommodate when considering how slow some of them were to recover from their wounds or to integrate into our social groups—but that average more than doubled by the early 1990s. We could scarcely build cages fast enough to house the new arrivals, and we spent every single day feeding and treating and ministering to our chimps, so much so that neither Dave nor I could ever get away from the farm. We even sold some more cattle to pay the bills. We were exhausted, but we never resented Chimfunshi or the chimps. Everything we did was done because we wanted to, and when you are helping animals, there is nothing to resent. The only thing that made us angry was knowing that other human beings were responsible for the war and suffering that had orphaned so many chimpanzees.

Food was becoming a problem, however. Our dairy herd was sufficient to supply each chimp a gallon of milk per day, but not even the local farmers could keep up with our demand for fruit and vegetables. Luckily, we were soon blessed by the opening of a new food shop in Chingola called Shoprite Checkers. For years, Zambia's stores had been poorly stocked and overpriced, and it was not uncommon to find food on the shelves that was out-of-date, spoiled, or riddled with cobwebs and weevils. But Shoprite Checkers, which is owned by a South African company, promised a stricter code of freshness.

One afternoon, a truck driven by Chingola's health officer pulled into the compound, piled high with apples and oranges, papaws and pears, cabbages and bread rolls and cream buns, and he asked if we could use the food for our chimps. Shoprite Checkers had planned to throw the goods away, since they were either bruised or out-of-date, but the health officer had been out to the farm a few times and immediately thought of us. Of course, I said yes, and we've continued to gratefully receive Shoprite Checkers' refuse ever since. It had always been extremely hard to get apples in Zambia, but the chimps soon made it clear that apples were their favorites—after cream buns, of course.

Meanwhile, the chimps continued to escape with frightening ease and frequency, especially once those inside The Wall got a look at their counterparts across the road in the fourteen-acre enclosure. Chimps are extremely territorial by nature and react violently if they believe their world is being threatened, and almost every escape devolved into a shouting match between Charley's group in the seven-acre enclosure and Chiquito's group in the fourteen-acre area, often with chaotic results.

There was one particular week I'll never forget. It was a terrible, draining, emotional roller coaster that left us all exhausted.

It began one morning when someone in the compound shouted, "Chimps out!" I ran to the seven-acre handling facilities and saw three chimps walking on top of the wall. Patrick ran around shouting at the chimps and they scrambled back into the enclosure. We searched all around The Wall looking for a hole, then went down to the river to see if they had got out that way, but could find nothing. The fact that an escape route was not obvious made us more anxious, and we resolved to watch the chimps more closely. But later that same afternoon, there was another shout: "More chimps out!" It was a repeat performance of the morning episode, with Patrick again able to scare the chimps back inside.

But instead of leaving to go check The Wall for holes, Patrick crouched down and watched the chimps for a while from the viewing platform, and saw Rita rustle about in the undergrowth, then pull a long, thick branch about eight feet long out of the brush. She looked furtively around, then propped the branch against the twelve-foot wall—and used it like a ladder to climb out. We called the bigger chimps into their cages, then went inside and removed the branch and a couple more we found lying near the same place. But if the chimps' ability to fashion a ladder was surprising, what was really amazing was that they chose the only spot around the six hundred meters of the wall that had a slight mound in front of it—making the wall at that juncture just nine and a half feet high. Even the chimps knew that an extra half-foot made a lot of difference. With the sticks removed, we relaxed again, assuming the chimps would find it difficult to find another ladder.

It wasn't even twenty-four hours later that the call again went up—"Chimps out!"—and then another more chilling shout: "Charley's out!" At once, people seemed to be running in all directions, but soon Charley came round a corner toward the house, followed by Little Jane, Rita, and Coco. We had recently received four dogs from the SPCA in Kitwe, and they were all looking confused and barking at these strange creatures. Charley rushed toward a couple of them and with a flick of his arms seemed to shoo them away, though not in an angry manner. While Dave and I were discussing what great difficulty we would have anesthetizing Charley, the chimp went to the refrigerator and helped himself to two beers—which we thought might make him mellow. But two beers did not seem very many, so we looked in our cupboard to see if there was any other liquor we could tempt him with. We found a half-bottle of cherry brandy that had been there for a couple of years. I brought the bottle outside and found Charley up an orange tree. So I pretended to take a drink and made lots of happy enjoyment noises, then offered the bottle to Charley. He took it from me and took a sip. He seemed to run the liquid around his mouth for a moment, then swallowed it, seeming a bit perplexed. Looking me straight in the eye, he took another sip, ran it around his mouth again, swallowed it, then shook his head in a clear "no" gesture and handed me the bottle back. Cherry brandy was obviously not his tipple.

An hour went past with me just following Charley around to make sure he did not get up to too much mischief. But then he wandered up to the electric fence around the fourteen-acre enclosure and started shouting at the chimps inside, who were making horrible angry noises right back, then upped the ante by throwing things at him. Offended, Charley took a step back, made a run at the fence, grabbed a pole, and in one mighty leap bounded over.

It was pandemonium. Milla screamed and fled toward the trees, followed closely by Maggie and Lucy. Noel stood her ground for a short while, but then ran for it, too. Choco and Leben, who had literally clung to each other since they'd arrived, instinctively turned and ran for the cages, hampering each other because they kept tripping and stumbling over one another's feet. They must have thought Charley was chasing them because their screaming got extremely loud—and when they eventually untangled themselves from each other, they ran off in separate directions. Sandy and Toby were left, but neither made an attempt to get away. In fact, what followed was amazing. Charley, who looked twice his normal size with his hair all standing on end, rushed over to Sandy and got him in a big hug while making loud greeting noises. Then he turned to Toby, whom he had never had contact with before, and greeted him like a long-lost friend. Charley hugged him, mouthed his head and shoulders, kissed him, and patted him, all the while making loud greeting noises. Toby never attempted to get away from Charley and in fact started making his own funny greeting noises. They moved apart from one another, but Toby went back to Charley and the greeting started all over again. This happened three times, then Toby slowly got into a cage and Charley followed him. Seeing my chance, I rushed in and slammed the door. This cage led to other cages, so after Toby and Charley had played around for about five minutes and Toby went through the slide door into the next cage, I was able to lock the door before Charley got through. At least we had him safe where he could do no harm to anybody.

While Charley's drama unfolded, the three girls had been roaming around, frightening the hell out of anybody who dared show their face. One of our staff fled down to the river, and Rita must have gone down there after him. I do not know what happened next, but I think the man must have hit Rita with his machete or at least taken a swing at her, because we suddenly heard that Rita was attacking him. Rita has never been nasty to any of our staff, but she was furious with this man! She knocked him over, bit his leg, and was trying to do him further injury when I managed to rush in and stop her. I did my best to calm Rita down and led her away, but then she saw the man again about ten minutes later and flew into another rage, chasing him up the path toward the house. Luckily, the man managed to get inside the house before she caught him.

It was about two o'clock in the afternoon by the time we got Charley locked in, and still none of the chimps had been fed. We went over to the seven-acre enclosure and called the chimps in for food, and though it took a long time, we eventually got everybody into cages, with the doors securely fastened. Dave and I then rounded up a gang of workers and went into the enclosure and spent the rest of the day removing any sticks, stones, or anything we thought they could use as tools. It was a lot of stuff, but by five o'clock, we were satisfied and let them all back out again. Charley had to be left in his cage at the fourteen-acre enclosure until we could figure out a way to safely move him back.

Early the next morning, just like clockwork, the familiar shout went up—"Chimps out!"—and my heart sank. Sure enough, there were Big Jane, Little Jane, Coco, and Cora walking along the top of the wall. By using their favorite food and a bagful of peanuts as bait, we got the chimps into their cages, and decided that was where they would stay until we got something sorted out. So we assembled another group of men and resolved to scour the enclosure, which is never easy, because the undergrowth is thick and almost impassable to humans, but we persevered and uncovered hundreds of potential ladders. Still, there were so many that removing them would have been an impossible job, so we decided that the best and quickest solution was to set fire to them. We'd had no rain for seven months, so the brush was quite dry, and once the fire started, it went very quickly. But the job of clearing and cutting all of the poles took another three days, during which time the chimps were kept in their cages. They were not happy, of course. Fights broke out and tempers were running extremely high, possibly because they could see the fire and were afraid; certainly they could smell it. Finally, on the afternoon of the third day, we let the chimps out and they all rushed forth, then spent the next few hours going round and round the seven acres in a most methodical manner, as if looking for something that they'd lost. The enclosure looked very strange from above, with all the black patches where we'd burned the brush, but the rains were imminent and would eventually make everything green again.

The next morning, there was yet another shout, this time from Patrick—"Chimps out!"—and all I could think was, "This is getting ridiculous!" Same scenario, different chimps, so once again we coaxed them into the cages with treats and went inside the enclosure. Sure enough, there was a telltale pole leaning up against the wall; where they got it from I'll never know. Dave decided he'd had enough, and announced there on the spot that the only way to keep the chimps in was to run another strand of electrical wiring along the top of the wall. The chimps were kept in for another two days this time—during which the same tempers and tantrums flared—but Dave and his work crew were exceptionally fast, and by the time the chimps were let out again, you could see they knew that The Wall was different. In fact, the escapes stopped promptly on the spot, and we all caught a much-needed rest while the chimps took some time to find another way out.

One day in May 1993, Dave and I were awakened at 1:30 A.M. by the telltale noise chimps make when there is something happening that they do not understand. It is almost like a slow, wailing, "*hooo, haaa,*" but very, very loud. We investigated the chimps' enclosures in the dark but could not find what was wrong, so we went back to bed and decided to look again in the morning. I'd gotten up as usual at 5:30 A.M. to begin preparing the chimps' breakfast when all of a sudden the dogs started barking madly. I ran outside and headed in the direction of the commotion, then suddenly stopped in my tracks. There, on the other side of the gate, stood Charley and a half-dozen other chimps from The Wall. I walked toward them and opened the gate, and Big Jane rushed up to me and threw her arms around me in a reassuring hug. She then patted my head before walking on, and Charley, Josephine, and Pal were next, each offering loud, happy greetings as they passed, and Pal stopped long enough to take my hand and give it a soft squeeze. I must admit, I was in shock.

As I walked unsteadily toward the enclosure, down the road came more chimps. Liza, Girly, Coco, Tara, and Boo Boo all sauntered past, a few shouting greetings or waving, and I held my breath as I neared the cages. When I turned the corner, I could clearly see the problem: The Wall had collapsed!

Apparently, the heavy rains had created a small ditch underneath The Wall's foundation, which grew weaker and weaker as it filled with water. When The Wall finally gave way, three panels that spanned approximately thirty yards collapsed and fell into the enclosure, creating a gap large enough to march an army through. Luckily, Spencer and Tobar were locked in a cage, and Cora, Bella, Little Jane, Donna, and Rita were still in the enclosure, but every other chimp was on the loose. I went back to the house to tell Dave the news, and we spent the next half hour making sure all the doors to the house, bedrooms, and cars were locked and all the windows closed tight.

I went back to the enclosure to see if I could encourage the chimps to give themselves up, and, to my surprise, I did manage to get Cora and Tara into a cage. Sometime during the morning, Dave reported seeing an unidentified chimp climb the outside of the electric fence and jump into the fourteen-acre enclosure, so we had that problem to deal with as well. But by tempting them with food and other treats, we managed during the course of the day to get Big Jane, Donna, Rita, Bella, and Boo Boo into one cage, which gave us nine down with seven to go. Unfortunately, the fugitives included two of the toughest chimps to deal with out in the open: Charley and Pal.

After much negotiation, I managed to entice Pal toward one of the cages, and I stood nearby to close the door as he climbed in. But when I started to shut the door, he suddenly reached out, grabbed my arm, and flung me across the room. My head struck the concrete wall and I sank into a heap. As I staggered to my feet, Pal stormed off. Later, Patrick and I were trying to coax Charley and Josephine into a cage at the fourteen-acre enclosure, when Charley also turned violent. He grabbed Patrick's leg and pulled it out from under him, slamming Patrick to the floor. The chimp next got hold of me and threw me down on top of Patrick, and again my head hit the concrete floor. Charley then ran off.

Woozy and aching from head to toe, I went over to The Wall, and apart from the chimps in their cages, I could see no others about. So I went into the feeding hallway between the cages and was trying to sort things out when into the passage stepped Pal. I had my hand on the door of a cage, and he must have thought I was going to try and shut him in again, because he immediately lost his temper. He raced forward, grabbed the front of my T-shirt, and tried to pull me down, but the T-shirt ripped down the middle. He made another grab and this time managed again to fling me around and down onto the floor like a rag doll, causing my head to smash into the concrete floor for the third time. I saw hundreds of pretty stars and promptly blacked out.

I do not know how long I lay there, but when I came to, Pal was sitting quite close by, just watching me. I staggered out of the passage and collapsed onto the grass outside, and there sat Pal again, watching. I believe he was feeling sorry for what he had done. I stumbled toward the house, and he followed me a short way, then sat down and glumly wrapped his arms around himself. Pal looked quite sad and confused, but I must admit I was in no condition to comfort him. This was the first time either Patrick or I had really been attacked, and though it was not as bad as it could have been, we were both shaken by how quickly and aggressively our chimps behaved. I had a huge lump on the back of my head and a splitting headache that would stay with me for about a week, but luckily, Patrick wasn't hurt at all.

About half an hour after Pal threw me, we coaxed him into a cage and he was all lovey-dovey again. At 4:45 P.M. we got Charley, who was pushing Josephine ahead of him, to walk through an open gate into the fourteen-acre enclosure. From there, we easily got him into one of the cages and closed the door behind him—then sat back and heaved a big sigh of relief. Liza and Girly spent the night in nests they had built in the big avocado pear tree in front of the house, and Little Jane slept in a tree near the broken wall, but the most difficult part of the ordeal was over.

Poor Dave spent most of the next day assessing the damage and wondering how he could repair it. He hired four bricklayers and four laborers and, although only three panels had collapsed, Dave was forced to knock down another three panels in order to strengthen the wall when he rebuilt it—a mammoth task. The project took nine days—nine days with the chimps locked in their cages. We were amazed at how well they behaved during that time.

The escapes continued at a furious pace over the next few months, and nothing we tried seemed to make much difference. If we patched The Wall, the chimps made new holes. If we strung electrical wire along the top of The Wall, the chimps laid branches over the wire and climbed out that way. At one point, there were even signs that one of the chimps had begun to dig what looked like a tunnel underneath The Wall. But digging straight through the twelve-inch bricks was always the fastest way out, and one day someone made a hole big enough for Charley to squeeze through. Fortunately, Pal, Spencer, and Tobar were in their cage together, so they did not get out, but all the others did. Most of the chimps got as far as the main gate to the compound, then changed their minds about going any farther. Still, it was a bad day for a breakout because we had a big group of people who were coming to see the chimps, and we were forced to post a lookout on the main road to stop them. They settled for a picnic in the bush about a mile up the road, and eventually left without seeing any chimps.

It took us most of the day, but we persuaded some of the chimps to go back into the enclosure through the hole in The Wall—which we then patched up—but by late afternoon, there were still seven chimps on the loose, and they were being shouted and jeered at by the chimps in the fourteen-acre enclosure. We became worried that Charley or someone else might try to get over the electric fence. Dave, Patrick, and I were exhausted by the constant breakouts and the ensuing battles to get the chimps back into their cages, so we decided to try something new. I'd wondered aloud if the noise of a shotgun being fired would frighten them back into the enclosure, so Dave went off for his gun.

Charley, Liza, Girly, Big Jane, Little Jane, Coco, and Tara were on one side of the road, facing the electric fence. Chiquito, Choco, Leben, Cleo, and all The Babies were on the other side, behind the fence. The two groups were fairly quiet but they were glaring very intently at each other and you could feel the tension mounting. Dave came back from the house with his gun and walked up the road between the two groups. I had hoped that the mere sight of the gun would frighten Charley's group away, since most of our chimps knew what a gun was and usually fled the second one appeared. But we had no such luck on this day; the stakes were already too high.

Dave fired one shot into the air and then another. What happened next was amazing—Chiquito's group behind the electric fence all stood up at once with their hair bristling and began shouting and waving angrily at Charley's mob. The noise was horrendous, and caused Charley and company to run screaming back toward their cages. It took about half an hour to get them all safely back where they belonged, but when I returned to the main road, Chiquito's chimps were still hugging one another and laughing and jabbering all at once. I believe they were congratulating each other on a job well done.

Ten

Billy

Billy was no bigger than a small dog when she arrived.

Of all the orphans to arrive at Chimfunshi, perhaps none has made a greater impression on me than Billy. That's because Billy weighs over fifteen hundred pounds, swallows entire cabbages in a single gulp, and still tries to sleep on the living room sofa. Billy is a hippopotamus we adopted in early 1992, and while raising a baby chimp is not that much different from raising a human child, a hippo is something altogether different. But love, perseverance, and creativity have combined to make Billy a very special member of the Chimfunshi menagerie.

It all started on January 8, 1992, when our son Tony came home from town with the news that some people had found a baby hippo whose mother had been killed. It seems local hunters had gotten an adult female down by the river, but were unable to collect the carcass because a bull hippo was still in the area, chasing people away from the body. When a tiny baby was discovered hiding under the dead mother, three local wildlife experts—Craig Wright, Billy Williams, and Colin Young—went down and managed to spirit the baby away. The little hippo was no more than ten days old, but the crowds at the riverside had nevertheless managed to inflict a number of wounds on its body by poking at it with sharp sticks. Tony thought perhaps they could do with some advice or help.

As he related this story, it occurred to me that Margaret Cook, our dear American friend, had once given us feeding schedules for various zoo animals. I dug the charts out of my files, and sure enough, one was for hand-rearing a baby hippo. Dave suggested I take the information over and try to help, but I was afraid Craig and the others might think I was interfering.

Dave insisted. "Sheila," he said. "We've got all these books. We might as well use them."

I made the two-hour drive to Craig's farm just outside Kitwe, and when I got there he rushed up to me and said, "Oh, thank God you've come, Sheila." I followed him into the house, and there, in his mum's bathroom, was one of the most adorable and helpless animals I have ever seen. The baby hippo was about the size of an average dog, with big brown eyes that had long eyelashes and a stubby little snout that curled up at the edges, as if it were smiling. Its back and legs were dark brown, but the extremities—its nose and ears and especially its toes—were bright, bright pink. For me, it was love at first sight.

They had the baby hippo in the shower, and the only way they could get it to drink anything was by pouring liquid on the tile floor and having the hippo lick it up. Every time they tried to get the baby to drink from a bottle, it refused. They had spent most of the night giving the hippo a rehydrating formula and some water, and it apparently consumed quite a lot, but there was no denying the baby seemed sad and lethargic. Meanwhile, the wounds on the animal were ghastly. There were about ten gashes on its back alone—three were fairly deep and wide open, the others not so bad. But the one on the underside of its belly just in front of its right back leg was the worst of all, and was so deep and so wide that Craig was able to insert four of his fingers and part of his palm into it. Being able to feel around in this big hole, Craig was able to ascertain that the lining cavity of the abdomen—the peritoneum, which supports the abdominal viscera—was intact, which was a lucky stroke for sure. But the risk of infection was still great, which made placing the baby in water impossible. Craig cleaned its wounds a few times with a sterile solution and then inserted some tetracycline powder, and the wounds began to close up in just a few hours.

By the time I got there, the hippo was no longer dehydrated, but it was clearly a bit lost. It had a name—Billy, chosen after one of his rescuers—and a sex: male. The hippo was also in need of a home. I said I would take him, so we loaded the tiny hippo into a big crate and placed it on the bed of Craig's truck. Because we had so many army roadblocks in those days and we were afraid the hippo would make noise and attract undue attention, I climbed into the box with Billy and pulled a tarpaulin over the top. We weren't doing anything illegal, but you never wanted to give the army a reason to be suspicious back then, and a woman sitting with a hippo in a box on the back of a truck would have seemed odd. It was best just to cover us both up.

Now, chimps are one thing, but a hippo had never been in our plans for Chimfunshi. Yet as I rode home in the crate alongside this tiny baby, it became clear to me that we couldn't just turn it loose. To begin with, a hippo is nothing but crocodile meat until it's about three years old, since it really can't survive without its mother, so simply turning Billy out into the Kafue River would've meant sure death. We had no choice but to keep Billy around the house until he was old enough to look out for himself, a task that we knew might well take years.

We reached the main compound about 5:30 P.M., then offloaded the baby hippo's crate onto our kitchen verandah and turned it on its side to get him out. But he decided to lie down instead, so the box became his home. I prepared a milk formula based on the zoo feeding charts, but putting milk in a bottle and getting a baby hippo to suckle on it are two entirely different propositions. In that first, long night, I tried every possible way to get some milk into Billy, all to no avail. I used baby bottles with baby teats, calf bottles with calf teats, I tried dishes, buckets, the floor, and at last a syringe with the needle removed. I'd get lots of milk into his mouth many times, then hold his head up to make the fluid go down. But the results were always the same: Every time Billy lowered his head, out poured the milk. Before long, I was tearing my hair out in desperation.

At about 7:30 A.M. the following morning, Dave came up with the idea that a shallow bath made out of a tarpaulin draped around an old window frame might help. He reasoned that since mother hippos normally nurse their babies in the water, perhaps that was what was keeping Billy from taking the teat. And if we kept the bath scrupulously clean, we figured, it would not infect his wound. So Dave quickly built a six-by-thirteen-foot "tub" on the kitchen verandah next to Billy's box and filled it about six inches deep with fresh water. Then we placed a canvas sack under Billy's tummy and lifted him up and into the water. Billy took a second to gauge his new surroundings, then snorted a few times, put his head down and drank a lot of the water, took a couple of steps, and defecated this sloppy yellow mess into the nice clean water. We got him out in a hurry and changed the water.

When we put him back into his pool an hour or so later, he again drank a lot of water. He also drank a lot later that afternoon. But at no time would Billy take the teat.

I was becoming desperate. By my calculation, Billy had not had any milk for at least three days, and I was sure he would soon become too weak to live. He weighed only about twenty-two pounds; how much longer could he go on? The fact that I'd been up all night didn't help either, and each successive failure hit me harder and harder. At one point, I was so deliriously tired that I was sure Billy was dying. We'd been battling and battling for almost twenty-four hours to get this poor baby to just swallow a little milk, but it was clear we were doing something wrong. My heart was breaking.

Finally, at about 4 P.M. I said to Dave that if we could keep Billy's head up when we placed him in the pool and stop him from drinking any more water, maybe I could get the milk into him. So we called Patrick over to help, and as David and I lifted Billy into the water, Patrick held a sack under the hippo's chin to try and stop his head from going down. I then scurried around and slipped the teat into his mouth. Billy battled and shook his head for a moment, then suddenly started sucking like crazy. In fact, in just a few seconds, he had drunk seven ounces. We all cheered! The sack-under-the-chin trick continued to work for the next few days until Billy learned to raise his own head for the teat, but by then the baby hippo who'd refused to eat had become downright voracious.

We also learned another vital piece of information about Billy as we fought that day—"he" was a "she." Determining the sex of a hippo at that age is tricky for anybody, and I'm sure Craig and the others weren't very concerned with whether Billy was a boy or a girl when they rescued her. Certainly I hadn't bothered to check in the first day and a half she'd been at Chimfunshi. But once we discovered the truth, we were faced with a dilemma: Should we change Billy's name? We tried calling her Wilhemina for a bit, even shortening it to "Willy," since that's not so different from "Billy," but nothing seemed right. Billy had been named after one of her rescuers, and somehow that name was the most appropriate. Billy she was.

Once Billy got accustomed to the bottle, she began to grow at an astounding rate. Three months after she arrived, I borrowed a tape measure that is normally used to gauge the size and weight of pigs, and Billy came in at 330 pounds. Three months after that, she was at 400. In no time she was eating five or six meals a day—a syrupy mixture of milk and porridge, spiked with vitamins—and began to follow me around like a dog, wagging her stubby tail whenever she sensed another meal in the offing. Years earlier, Dave had built a small swimming pool for our grandchildren on the bluff next to our verandah, and this quickly became Billy's private bath, since she needed to get used to swimming and we didn't dare take her down to the river's edge yet.

The best times with Billy were in the evening. While Dave and I sat back and enjoyed a beer, Billy would splash about in her pool, sometimes bobbing up and down for over an hour, then look around to see what sort of mischief she could get into. If a peacock passed too close to the pool, Billy would spring out of the water and chase it around the lawn. She liked it when we joined in, and would chase us around, too. She'd sometimes run too fast and slip on the damp grass, but then be up again and off after something else. When she tired of that game, Billy would either go back into her pool or come and sit with us at the table, sighing contentedly as we stroked her head. By about six-thirty, she was ready for bed, and would plod off into her kennel for the night.

We did have one very bad night with her, though. I'd gone to give Billy her last bottle of the day and found both her and her enclosure overrun with army ants. Billy was rolling around, trying to get the horrible, biting things off her, and when I opened the gate, she got up and shot past me to her pool. We eventually got all the ants off her, but realized that she could not sleep in her own enclosure until we killed the ants. After trying a few places around the compound, we finally ended up letting Billy into our lounge. Billy went in easily enough, but refused to stay unless we stayed with her. There was an old couch in there, which I opted to sleep on for the night. But every time I moved she got up and touched my head to see what I was doing. So I put the cushions on the floor and crawled onto them. She lay down next to me, and we both had a very good night's sleep. I have slept with a lot of different animals in my day—some of them cuddly, some of them not—but it is very difficult to cuddle anything that weighs four hundred pounds.

Billy quickly came to regard me as a mother figure, and would come trotting around the corner of the house at the mere sound of my voice. But she also developed close, if strange, relationships with the other animals at Chimfunshi, particularly the chimps. I think it was just that she was so lonely with no other hippos about, and the chimps were noisy and rambunctious and she was attracted to that. I think she felt she was with others of her kind, even if they looked nothing like her. Of course, most of the older chimps were terrified of Billy, and would scream loudly whenever she got too close. But we had a group of babies at the time that we called The Infants, and they lived in the nursery cages next to the house. That group—Trixie, Diana, Doc, Zsabu, and Violet—all required daily feeding and constant human interaction, which meant that Billy was always close at hand. Billy came to adopt those chimps as her own, and would spend every afternoon either napping up against their cage or gazing happily at them through the wire mesh. I used to say she was "baby-sitting," because she would get up every now and again and look at them very carefully, as if counting to see that they were all there. The Infants grew quite comfortable with Billy, and would take turns reaching through the wire to slap her hide or pull her ears, and on bush walks in the orchards, they'd jump down from trees onto her back and play all over her. She seemed to love it.

When we opened the fourteen-acre enclosure, we decided to build a new nursery for the baby chimps and a food storage hut close by, in order that The Infants might see and hear the older chimps and begin the process of socialization. But when we moved The Infants, we neglected to inform Billy, and when she waddled up to their old cage one morning and found it empty, she was utterly lost. Billy refused to go into her room after her breakfast bottle, which she normally drank in her pool, and insisted on following me everywhere I went. This proved very difficult, because there were places I did not want Billy to go, but she insisted. She forced her way into the kitchen, then followed me out when I left. She got into the food storage hut and ate a week's supply of sweet potatoes, then left peacefully when I shooed her out. This went on all morning until I got near The Infants in their new cage and Billy, who was no more than two steps behind me, suddenly stopped dead in her tracks. She trotted up to the cage, rested her head against the bars, sighed loudly, and sat down to gaze at them—glad to have found them at last. The Infants all came over and gave her a pat on the snout—or, in Zsabu's case, a punch—and Billy seemed much happier.

Billy also became attached to many of our dogs, and began to mimic them in her daily actions. If they were playing with a ball or a stick, she would grab an automobile tire and fling it around like a chew toy. If they growled at strangers, she, too, stood on guard. And if the dogs flopped down for naps in the shade, then Billy laid herself out right alongside. Her closest companion was Gretchen, a rottweiler we'd had for many years, who often slept with Billy at night. But I knew something was up one morning when I went to feed Billy at six o'clock and she refused her bottle, pushing me instead toward the kennel, where I found Gretchen's body. It seems the dog had died during the night, and Billy was inconsolable. I sat with Gretchen for a while, sort of saying a last farewell, but Billy kept nudging me as if trying to make me do something to revive Gretchen. She refused all her milk and food for two days and kept a silent vigil near the kennel. It was during this time that she started to get into our bedroom at night. She got in three times—wreaking havoc on each trip, as she pushed down the door and crashed around our dressing table—so Dave decided to try and keep her away by stringing electric wire around the house. After getting shocked a few times, she eventually learned better and stayed outside—for a while, at least.

Billy did become an excellent watchdog. On more than one occasion she stood her ground during a chimp escape and chased the offenders away from the kitchen, and once kept Charley and Josephine up a tree for more than an hour. Another time she actually rushed to the gate to head off a chimp assault on the house, chasing the chimps back in the direction from which they came. She weighed well over five hundred pounds by that time and was the size of a small car, so even the toughest chimps were terrified of her. Billy's diet had grown to include three large buckets of vegetables each day—along with her usual bottles of milk mixed with porridge—and yet she still behaved like a lapdog: If I sat down somewhere, Billy would invariably come over and place her head on my knee.

One morning, Billy and one of our mongrel dogs, Sam, were behaving very oddly when I went outside to begin my chores. Sam kept barking wildly and running away from me toward some baboon cages about a hundred yards from the house, while Billy was pushing me in the same direction, using her big snout to move me along. There did not seem to be any other option but to go where they wanted me to—down the path and through the bamboo grove to a lemon tree in front of the largest baboon enclosure. There, both Sam and Billy were acting in a threatening manner at the base of the tree. At first I could see nothing, but hidden under the leaves was a gaboon viper—one of the deadliest snakes in all of Africa. I called Dave, and he had no option but to shoot it with a rifle, because Sam and Billy would not otherwise have left the spot. I have wondered since if they were trying to protect the young baboons who were nearby in their cage. Or if perhaps they were trying to protect me.

Billy was just one of the many animals who took up residence at the orphanage in those days. Someone once gave us an orphaned bush baby, a small, squirrel-like animal with huge eyes, and asked that we return it to the wild. Bush babies are nocturnal animals with tremendous leaping abilities, but I'm afraid their huge eyes make them appear to be perpetually in shock. This bush baby, whom we christened Leon, was only about five months old at the time, so we decided to let him gain some size and strength around the farm before being let loose. During the day, Leon slept in a traveling box in our lounge, and we'd only wake him around five o'clock in the afternoon to give him his breakfast—usually a meal consisting of milk, ground chicken, peanuts, porridge, and a banana. After eating, Leon would bound out the window and away into the trees, and we would not see him again until the next morning, when he would magically reappear, sound asleep in his box. But one day, Leon did not return to his box, and our hope was that he found some of his own kind to live with.

We've received over a dozen African gray parrots from game rangers over the years, and they live in a large flight cage just outside our verandah. But one, Smokey, was so tame that he lived with us in the house. Smokey ate off our plates and drank anything we drank, including tea as hot as you could stand it. The one thing Smokey would not eat was brussels sprouts. If he found them on the plate, he would grab the offending vegetable in his bill and fling it away over his shoulder.

Like all parrots, African grays can grow to be quite old, and another of ours, Eddy, must now be almost forty-five years of age. He once developed a crush on another African gray who was at least thirty years old herself, and flirted with her shamelessly. We never expected anything to come of it, but one morning we found she had laid an egg on the floor in a very dark corner of the cage. In all the years we'd had parrots, this was the first one who had ever laid an egg. I thought perhaps the parrot would like a nest box, but when I opened the door to the enclosure to see what I could do, Eddy flew at me and bit my foot, then chased me out, the female joining in by making very loud growling noises. I managed to put some soft grass near her to make a nest, but she didn't seem the least bit interested. A few days later she laid two more eggs, and then another the day after that. The parrot sat determinedly on those four eggs, even though they rested on the concrete floor, and would not touch any of the nesting material I laid nearby. I was very excited by the prospect of having baby parrots, especially a species as rare as an African gray, and counted down the usual twenty-day gestation period with great anticipation. Unfortunately, the eggs failed to hatch, and the parrot eventually lost interest in them.

Baboons and monkeys remained some of our most regular guests at Chimfunshi, and they were also among the most dangerous. Dave was once down in Lusaka on business when someone sent him a message asking if he'd mind collecting an adult vervet monkey and bringing it back with him to the orphanage. Dave agreed, and gave the owners a carrying box and told them he would collect the monkey the next morning. But when Dave arrived, he found that the owners had been unable to get the monkey into the cage. And unbeknownst to Dave, the caretaker had given the monkey a beer, hoping it would calm the poor animal down. It did just the opposite, however. The monkey was enraged, and scrambled away from its cage and attacked the only stranger there—Dave.

When a monkey bites, it doesn't just nip with its teeth; it grabs hold with its hands, sinks its teeth in, then pulls its head back, all in one terrible split second, ripping whatever it has bitten in the process. This monkey removed a piece of flesh about four inches in diameter from Dave's left forearm and an even bigger piece from behind his right knee. The ligaments, nerves, and even an artery were severed in Dave's arm, and blood was gushing out all over the room.

There was panic and pandemonium, but someone managed to get Dave's wounds bundled up in towels, then drove him at great speed to the University of Zambia Teaching Hospital, a distance of approximately twelve miles. It took half an hour before they got Dave into the operating theater to stitch him up and stop the bleeding. With the threat of AIDS, nobody in Africa is happy about receiving transfusions nowadays, so he chose not to accept blood during the operation. But Dave had lost a great deal of his own blood during the ordeal, and he was very weak and sick for a long time afterward.

I got a message over the wireless radio from Lusaka about Dave's injury the same afternoon, and managed to get a lift in a small plane the next morning. When I arrived, Dave looked dreadful and was still in tremendous pain, but I was only able to stay with him for two days—somebody had to be at the farm. Dave remained another week in Lusaka and came back by plane a week later. He visited our local surgeon in Chingola, but none of the painful skin grafts they attempted were successful. It took three months for Dave's wounds to heal up, but he still bears ugly scars from the attack and his left hand has not regained its full use because of the severed nerves.

Even though our objective at Chimfunshi is to rehabilitate as many animals as possible and return them to the wild, some seem to prefer our house to the great outdoors. That happened with Ole, a tiny little barred owl that was brought to us after it had fallen out of a nest during a thunderstorm. Ole was only six inches in length and was still too young to have feathers, but he had the most exquisite markings. His upper body had thin bars of brown back and forth, and his wings bore a row of very bold white spots that formed a V from his shoulders to his wing tips. Ole somehow got out of his cage the first night he was with us, and promptly claimed the entire living room as his personal domain. When he grew old enough to sprout feathers and eventually fly, Ole would swoop all around the room, dive-bombing our heads and our dinner plates.

Once, I was sitting at the dining room table enjoying a fried egg on toast when suddenly Ole swooped down and snatched it off my plate. I like my yolks soft, so this one dripped all over the table and down the curtain after Ole perched himself on the curtain rail. He ate quite a bit of it, then let the rest drop onto the carpet. He did the same with a piece of currant bread that Dave was eating—snatched it right off the plate, flew with it to the back of the settee, then broke it into little pieces, which scattered everywhere. With the bread, the egg, and the other presents he kept dropping, the house became an absolute mess. But we loved little Ole—he was especially endearing when he came and sat next to you in the evening and wanted his head scratched. He'd close his eyes and drift off, really seeming to enjoy it.

After Ole had been with us for about two months, we opened a window to see what would happen. He looked for a long time, then spent several minutes gathering up his courage before flying out through the window to a nearby tree. Ole sat there for about half an hour before taking off again, and he did not return for the next three nights. I missed him terribly and assumed he was gone for good, but the next morning I could hear him chattering and clicking his beak outside. I opened the window and he flew in right past me into the lounge. When I offered him some food, he ate ravenously. He stayed all day and flew off again that evening, and only rarely comes around anymore.

We've received a steady supply of bush buck over the years, all of which we've released onto the open ranges of our farm. A bush buck is a medium-sized antelope whose fur is a beautiful red-brown with white stripes on its back and white spots on its shoulders, hindquarters, and face. This is a lovely animal—its spots look like patches of sunlight when you see them at a distance. One particular bush buck, a female we called Bellie, was sweet and gentle and would follow me around the yard like a puppy whenever I did my chores. She also had an incredible sense of timing, and knew exactly when I would begin to prepare the chimps' food. She'd invariably be standing outside the food storage hut waiting for me, and grew particularly fond of sweet potatoes and bananas. When she eventually became pregnant and wandered off from the main compound to give birth, we did our best to stay away in the hope that her baby would grow up wild. But we often saw Bellie and her fawn at a distance, and if she came near the house, we felt certain the baby was hiding in the tall grass nearby.

One afternoon not much later, a man with a spear and three dogs was seen walking across the river upstream from our farm. Due to a drought that plagued Zambia in the early 1990s, the river had dropped to the lowest level we'd ever seen, and it was possible to simply walk across in places that normally required a boat. We did not see Bellie for over a week and were becoming very worried, but we were shocked when she did finally turn up. She had an enormous hole in her left side that was oozing thick, green fluid and was running down her stomach. She seemed very weak and had lost a lot of weight. She also appeared to be alone.

We offered her food and water and managed to hold her while we did what we could for the wound, then gave her an injection of a long-lasting antibiotic. After examining her udder, we realized she must still be feeding her baby, so we felt better about that, but we obviously could not confine her. She managed to feed her baby for the next four months, during which time the wound in her side got smaller and smaller. It eventually healed up and she began to put on weight again, but the experience had clearly left her spooked, and neither she nor her baby strayed far from the house for quite some time.

Billy, meanwhile, continued to grow at a staggering pace, and could surely have fended for herself in the Kafue River by the time her weight reached a ton and beyond. But she still insisted on squeezing her bulk into the tiny wading pool near the verandah, and nothing I did could induce her to set foot in the river. I tried twice a day for two weeks to get her to go into the Kafue, and we'd spend hours standing on the bank, alternately staring at each other and watching the water rush past. Every time I told her to go in, Billy just looked at me, as if waiting for me to go first. She'd even nudge me toward the water from time to time, as if she thought it was me who needed encouragement. But I was too afraid of the crocodiles, so I'd get into one our boats, hoping that Billy would follow me into the water. All Billy wanted to do, however, was climb into the boat with me.

Then, on July 8, 1994, Billy went into the Kafue River for the first time. I'm not quite sure how it happened, but she was standing alongside me on the jetty and just sort of slipped and fell in. What a wonderful experience that must have been for her, to suddenly have all of her bulk buoyed up by water! She got so excited that she was jumping out of the water like a porpoise—swimming long distances from the boat and rushing back to touch my hand and make sure everything was all right. Dave came down with the video camera and was able to capture this wonderful event until it was too dark to film anymore. From that point on, Billy spent every night in the river, splashing and wallowing about, then would return to the house each morning for her breakfast and a mid-morning nap just outside the main gate.

A few months after Billy's first dip in the river, I was awakened by a tremendous amount of noise in the middle of the night. The chimps in The Wall were going crazy, screaming and yelling hysterically, and the noise was so bad that I had to get up and go find out what the trouble was. I stood on the viewing platform for about ten minutes and peered out into the enclosure, but it was rather dark and I could not see much at first. Then I heard a rustling in the bushes below, and I could just make out something dark moving about in the shadows. Sam the dog came up the stairs to the platform and jumped on the wall to have a look for himself, but I was scared that he might jump in, so I shouted, "Get down, you silly dog!"

At that, the undergrowth exploded in a frenzy of snapping branches and rustling leaves, and Billy emerged, heading at full speed toward the cages. She must have gotten into the enclosure by way of the river, and now that she'd heard my voice, she was trying to get into the handling area. The doorway was far too small for her, but I was scared that if she pushed hard enough, she would get wedged into the opening—and with all those chimps in there, that would have been disastrous. Knowing I was there up on the platform, Billy would not go back and out by way of the river, so I had no option but to let her out by the big steel door in the side of the wall. Fortunately, it was dark and the chimps were scared of her, so I was not too worried about Charley and the others trying to get out as well. But it set a tiresome and frustrating precedent, as Billy soon came to regard the seven-acre enclosure as part of her regular route.

Almost every day, Billy would plod down to the river, swim downstream, then come ashore inside the chimps' enclosure, sending Charley and the rest into hysterics. When we'd call the chimps for lunch, it was Billy who would come waddling up the path instead, and the chimps would stay glued to their trees. Although they eventually got used to Billy's presence, they would still not venture near the handing cages if she was about, meaning that one of us (usually me) would have to go down to the river and call, "Billy, Billy, Billy, it's bottle time!" at the top of our lungs. She'd slowly meander down to the river and then up to the house for her bottle, the only enticement we found to lure her out of awkward situations.

Billy's desire to be with the chimps was surely driven by her loneliness, as were her attempts to break into the house. Night after night, I'd barricade chairs and table and pieces of iron and steel around the front and rear doors, yet somehow Billy would pick her way through, step over the electrical wire, shove down the door, and squeeze inside. Unfortunately, the dogs regarded her as one of their own and would not so much as bark, so Billy's break-ins often went undetected. Until the next morning, that is. Then I'd awake to a kitchen of half-eaten vegetables and overturned baskets, gnawed pillows and chewed-up rugs, and I'd know exactly what had gone on. Once, she must have tried to climb onto the sofa again, like she did when she was small, because I found it splintered into a thousand pieces. But it must be terrible to be the only one of your kind, and I'm sure Billy only broke into the house because she figured it was her home, too. After all, it was where she'd grown up.

As Billy got bigger and bigger, I began to fear that she might hurt someone. I was more worried about that than about any threat the chimps might have presented. We'd begun to get hundreds of visitors each week to Chimfunshi, and Billy soon became quite well known. For rather than hide from the crowds, she was attracted to them, and was particularly fond of small children, waddling along after them as they toured the farm. People never understand or believe that a hippo can sprint at thirty-five miles per hour or leap several feet into the air, and since she was only tame around people she knew well, we expended a lot of effort keeping her away from the tourists, and vice versa.

Billy's potential for causing injury was enormous. I know—twice she nearly caused me serious harm, and both times it was completely by accident. The first time, I was coming down the path from the chimps' enclosures and she was walking next to me, and all of sudden she put her foot on my foot. I don't know quite how it happened—probably I turned some way she wasn't expecting—but she was suddenly standing on my foot.

I shouted, "Billy!" more out of surprise than pain. And because I shouted, she froze—but in her freezing, of course, her foot remained on my foot. She weighed at least a ton in those days, but fortunately, a hippo's foot is soft, with little toes that stretch out, and her foot somehow spread out over the top of mine. It still hurt like crazy, however, and she turned around to look at me quizzically.

By that time, I was frantically pushing and prodding her to move, and eventually she figured out the problem and delicately lifted her foot. I thought she'd broken every bone in my foot, of course, but she hadn't. It was sore for a few days, but it was all right in the end.

I also fell under Billy once. There's a little gate in the fence down by our baboon cages, and I was walking ahead of Billy toward the gate, with her coming up behind me fairly fast. As I tried to step through the gate, my foot caught on the electrical wire stretched across the ground. Thankfully, it wasn't switched on, but it was enough to trip me, and I fell to the ground. She was moving too fast to stop, and as I lay there, I thought I was done for. As Billy landed on top of me, however, she froze again, managing somehow to keep her weight off me, even though she was poised right above me. I don't know how she did it, but Billy managed to slowly and carefully sidestep her way off me, then turned around and tried to help me get up, nudging me with her nose. It was incredible. All she'd needed to do was put one foot on my chest and I'd have been dead. I was so frightened, I almost stopped breathing.

Billy's insistence that she belonged right alongside people made her increasingly difficult to avoid. Once, one of the local priests from the St. Joseph's mission, Brother Tony Droll, came out to the farm with a visiting friar from the United States, Father Vince Peterson. They and a group of seminarians from Kenya wanted to see the chimps and have a picnic in the orchards, and we were happy to oblige. After lunch, they gathered under the avocado trees and began to celebrate Mass. Father Vince set up a portable altar, and, fully vested, began the Mass, with everybody else sitting on the grass in a circle. But at the beginning of the first reading, Billy came plodding through the undergrowth and walked straight into the celebration. There was a bit of squirming and uneasiness, but the seminarians, strong in their faith, remained seated in the circle. I'm sure none of them had ever been that close to a hippo before, and clearly all of them were thinking that if she stepped on them, they'd be dead.

During the scripture reading, Billy seemed content to just stand there, listening meditatively. Then, Father Vince began his homily, even though all ears and eyes by now were on Billy, who, perhaps bored with the sermon, put her head down, picked up someone's baseball cap, and began noisily chewing on it. A minute later, she spat out the tattered wad that had once been a hat and looked up, with new interest, at the altar cloth, then began to move toward the altar. When her nose was less than two feet away, Brother Tony picked up the altar—hosts, chalice, and all—and began moving it away as Father Vince continued preaching. But Billy was determined, and kept coming. Brother Tony moved the altar again, while Father Vince followed behind, continuing his sermon. Finally, in order to take Billy's attention off the altar cloth, we gave her an orange, then another, and another. When she'd sufficiently shifted her attention to eating the oranges, the entire celebration was moved about twenty-five yards away to the shade of a clump of bamboo, and Father Vincent continued his homily.

But Billy had her heart set on the altar cloth, because after she ate the oranges, she moved her huge bulk over to the new celebration site and once again began to edge toward the altar. Brother Tony thought that perhaps if he stood in her way, she would stop. Wrong! Billy came right up to him, lifted her slobbering lips to his stomach, and nudged him messily aside, as Father Vincent continued preaching. Brother Tony tried turning his back on her, but that was no better. When Billy again moved him aside, the altar was again moved as well.

Father Vincent began the prayer of the faithful, and he intoned, "Lord, grant that Billy will find her place this afternoon and we will find ours."

We put some more oranges on the ground and the seminarians added some of their leftover picnic items, and we moved the congregation and the altar back to the shade of the avocado tree. The unique celebration came to an end just as Billy was finishing her snack, but she decided to head off to the shade for a nap, rather than pester the visitors any further. Then, as people were getting into their vehicles to return home, Father Vincent asked the group, "Is there anybody who could give a short summary of my homily today?" There were confused looks and sheepish smiles among the seminarians, but nobody answered.

One night, after the summer rains had flooded the fields and raised the Kafue River to its highest point of the year, Billy failed to come home. She normally spent all night in the river and was once spotted two miles away, but she'd never failed to return for her breakfast bottle. She was gone for six days—six of the longest days of my life—and then came home on the seventh day and was her usual self again. Apparently, the last few wild hippos in the area had congregated downstream and Billy had gone off with them for awhile.

That established a pattern of roaming that is both comforting and terrifying: Billy will go off for days with the other hippos, presumably feeding and courting and behaving as she should, but we worry that her lack of fear of humans is liable to get her into trouble. A farmer who lives about two hours downriver by boat radioed over one evening to say that he'd spotted Billy outside his house with another hippo. He knew it was Billy because she opened his back gate and strolled around his generator house, then left without damaging anything.

Another time, Billy was gone for four days, and returned home with a few nasty cuts on her right side. One of these was in the shape of a question mark about fifteen inches long. Most of the cuts healed up very quickly, thank goodness, but the question-mark scar remains to this day. Hippos in the wild often injure one another with their long tusks during fights and courtship, and we wondered whether her wounds had resulted from the advancements of a male hippo or from a fight with another female. Either way, it was clear that we could no longer protect Billy like we used to.

Eleven

The Enemy

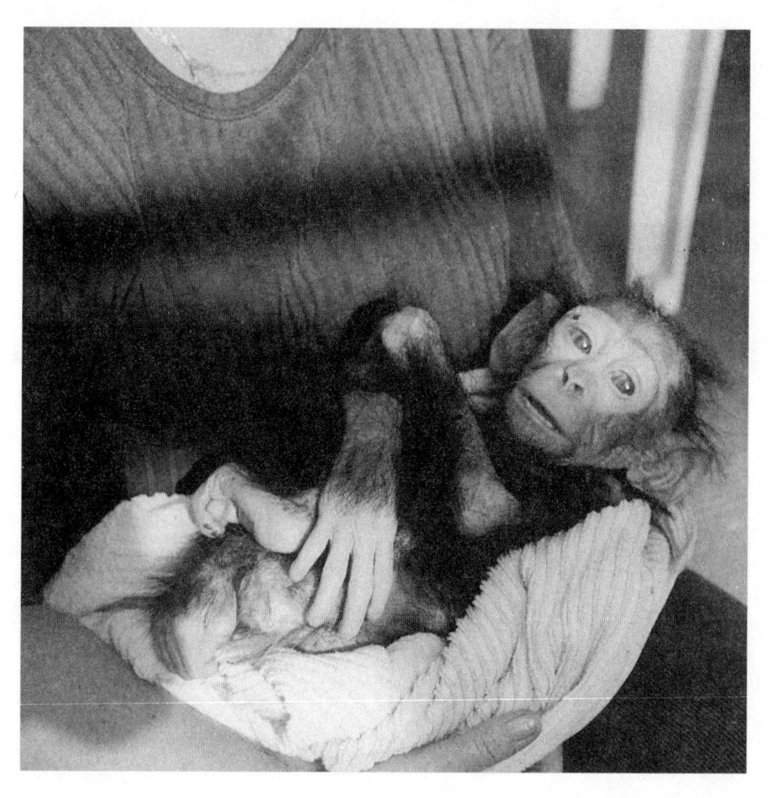

Clement was one of the many chimps that arrived more dead than alive.

Chimfunshi exists because mankind has made it necessary. In a world where human beings go hungry, AIDS and overpopulation remain unchecked, and forests are chopped down for profit, it's easy to understand why chimpanzees get such short shrift. But that doesn't mean I have to like it. Or accept it. As long as there are laws that protect chimpanzees—or any wild animals, for that matter—Dave and I are prepared to do our best to see that any animal that needs help or sanctuary can have it here. Scientists have shown that chimpanzees are extremely intelligent, social creatures with many of the same emotions as humans, and are blessed with a capacity for communication that we are only just beginning to comprehend. Yet many humans still hunt them, eat them, cage them, and abuse them, and it's a wonder any of the chimps we get at Chimfunshi ever recover from these ordeals. That they do says more for the resilience of chimpanzees than it does for the "humanity" of human beings.

But I struggle to even classify poachers and smugglers as human beings. To me, poaching is despicable. If you have to kill animals, there are humane ways of doing it. I went out hunting once in the 1950s with a German chap, who was a professional hunter, and we trekked out into the bush near Lusaka, back in the days when the Zambian capital was still quite small and the wildlife lived just on the fringe. The hunter said he needed to go out and get some meat, and I said I'd go along. I didn't carry a gun, of course; I simply wanted to try and understand what this need to destroy innocent creatures was all about. But he didn't kill a thing that day. Every time he saw a duiker or a bush buck or an antelope, he'd take a long look, shake his head, and say, "No, that one's still nursing. Must have a baby nearby," or "No, too young," or "No, the shot wouldn't be right; might wound it." I was impressed, he had a respect for the animals he meant to kill. But I'm afraid his kind of hunter is hard to come by anymore.

Today, it seems that all the hunters in Africa do is put out rusty wires or string snares, then leave them untended for a couple of days. They know that any animal caught by these traps will then sturggle to escape, with great suffering, but they don't care. That's the sort of poaching we get here on the farm. We hear the gunshots, or see the hunters crossing the river onto our property, or find their fires in our forestland. We've uncovered hundreds of snares over the years, using so much wire that we could refence our entire property with it if we chose to. In the old days, people used a rifle to get their food, and many of the locals would go out to kill an antelope for that week's meat. But that was subsistence. Nowadays, it's people going out and killing everything in front of them to get the hides or trophies to sell. There's no thought of tomorrow for poachers today.

One day in 1992, I drove into Kitwe with Carole Noon and Nikki Ashley, the former head of the Zambia Wildlife and Conservation Society. Nikki had not eaten, so we decided to go to The Cobweb, a shop that sells local curios and does a nice tea, but as we approached the store, a lady who knew me shouted from a passing car, "Sheila, quick! There is a man at The Cobweb trying to sell a chimpanzee!"

I hurriedly parked my truck and asked a security guard about a man selling a chimp, and a lady worker came out of the shop and said that he'd just left with another man. She'd given the man my name and address, she said, but before she and I could speak further, the guard shouted, "There he is!" and ran off down the road, calling after the two men.

The guard returned with two Zaireans, a pair of scruffy individuals who looked somewhat wary and confused. Unfortunately, neither of them could speak English. Nor could we speak French or their native dialect, so a few moments of utter chaos followed as I tried to find out more about the chimpanzee. Luckily, another local man overheard our conservation and offered himself as an interpreter. I put on a big show, saying, "Ooh, I want a chimp! I want a chimp! How much do you want for it?" And the man told me that the chimp was just a baby, that they were keeping it in the compound where they were staying; and that they wanted $2,500. In dealing with poachers and smugglers, I've learned that they will believe almost anything you say, as long as you are discussing money. So I said that I thought it was far too much, and anyway, I didn't have that kind of cash on me. But I added that I had a friend who might lend me some of the money, if they'd agree to accompany me to his house.

The two men—who did not appear to be armed, though I couldn't be sure—and the interpreter got into the back of my truck, and my mind raced as I tried to figure out what to do. I knew the most important thing was to get hold of the chimp;

once I had the infant safely in my arms, I could deal with whatever else came up. I drove slowly in order to plan things out, and luckily, the Zaireans did not appear to know that we were going in circles. I think the interpreter knew something was up, and I feared that he might say something to warn the others. But he chose not to say a word, even when we turned down the same road for the second or third time.

I drove to a garage whose owner I knew to be a game ranger, telling the two Zaireans that he was a friend who might lend me some money. In fact, I hoped he'd arrest them on the spot, but the ranger was not home, so I whispered quickly to his daughter and son-in-law to play along with me. I went through the whole Please-lend-me-money-I-want-to-buy-a-chimp routine, and the son-in-law was great. He pretended to be very angry with me.

"How can you expect me to lend you money when I can't even see the merchandise?" he shouted.

The interpreter relayed all this gravely to the Zaireans, and they actually seemed to understand the son-in-law's anger. We agreed on a price of $2,000, but only if we got to see the chimp first, so off went Carole and I to their compound, while Nikki and the people at the garage stayed behind and started phoning police and game rangers in the area.

We pulled up to this dismal little hut in the middle of a small village about three miles outside of town. The building had two rooms, both very, very dark, and upon entering the second room, I could just make out this pathetic little thing straining at the cloth strips tied around its legs and to the rungs at the top of the bed. When I saw that chimp, I was filled with not only rage, but a sense of nausea that left me dizzy. Yet I knew I had to keep up the act lest the Zaireans get suspicious. The very frightened baby grabbed hold of me, and I untied the knots that held it. I got the chimp into the car, and we all drove back to the garage.

As we arrived, I heard someone shout, "They're back! Close the gates!" But I only realized we were safe when a game ranger and two policemen pulled up and began questioning the Zaireans. After learning that the two men did not have any passports or transit papers, the policemen arrested them and we all went down to the station house. The police took about half an hour to decide what to charge the men with, but Nikki was very good at listing all the laws these men had broken—which was fortunate, since police in Zambia are not always very informed about conservation laws. But the men were in Zambia illegally, and that allowed the police to hold them until deciding what else to charge them with.

We drove straight home with the new baby. He was very thin, with almost no flesh on his bones, but he didn't appear to be too terribly dehydrated. He had badly swollen glands under his chin and a few marks at the top of his legs—probably wounds from his capture. However, apart from being very sad, frightened, and confused, he wasn't as bad off as one would have expected. We gave him a tremendous amount of care and attention—which he clearly loved—but the problem of his name arose. Carole was leaving us shortly to return to the U.S., and because it was around the time of her birthday, we decided to name the baby after her. Unfortunately, it was a boy, so the name "Carole" didn't seem quite right—nor did "Noon." So, since Carole was working toward her Ph.D., we named the chimp Doc.

I am not a violent person by nature, but we do keep loaded guns at the farm, and I've always believed I would use them on a human being long before I'd use them against an animal. Poachers are a constant threat, and both Dave and I are committed to protecting our animals. Though we protect our chimps as much as possible, they are, in a way, their own security: Because chimpanzees are extremely territorial, they regard the area inside their fenced enclosures as their property and will react violently if they feel it being threatened. But Dave and I are constantly on guard against people who would steal or harm our animals. Once I nearly shot a man whom I thought was trying to sneak into our farm. It was just about dusk and the light was failing, but I could see this figure climbing over the fence and I had a gun in my hand as I called out a warning. All of a sudden, I heard this familiar voice screaming, "Madame! Madame! Don't shoot! Don't shoot!" It was one of our drivers, who had started home for the day but had forgotten something and couldn't get through the gate to retrieve it.

But the time I truly realized my potential for violence was when some poachers got to my animals. We had these two geese that were very old, and they'd been with us for so many years that they couldn't have babies anymore. When someone gave us three little goslings, though, the female goose adopted the goslings and suddenly we had a family of geese. It was lovely; they'd parade around the farm in a line, and both of the elder geese wound up being excellent parents. But one morning I woke up and found they'd been stolen from the empty baboon cage where they'd made their home. I was so angry I felt sick. I sent two of our workers out and gave them each two bags of oranges, telling them to go across the river pretending to sell them while finding out anything they could. Meanwhile, I looked around the farm for clues.

Maybe an hour later, a local fellow I knew by sight came racing up on a bicycle into the main compound with the female goose under his arm. She looked a bit bedraggled but was alive, and I ran up to him and took her. Before I could ask any questions, he said breathlessly, "They've got them! Your boys have got the poachers! They're out on the floodplain, but they need transport!"

We had a Land Rover and a tractor-trailer in those days, but the Land Rover had a flat tire, so I thought it would be faster to take the tractor-trailer. I grabbed the shotgun and a billy club and drove onto the main road across the floodplain.

Out in the middle of the tall grass was a large crowd of people, so I steered in that direction. I also laid the shotgun down on the seat of the truck—which was fortunate, because of what happened next. The truck had barely come to a halt when I leaped out, still holding the billy club. I could see that the crowd had encircled two rough-looking men. Somebody had my other goose—also still alive—under his arm, but I could not see the goslings anywhere, so I shouted, "Where are the babies?"

Somebody thrust forth a cooking pot. When I looked inside and saw my three dead goslings, I can't tell you what came over me. I screamed and rushed at the two men with my billy club and was prepared to bash their skulls in—I really was—but a policeman jumped out of the crowd and grabbed me and shouted, "No! No! Madame! You can't do that!"

Nevertheless, the crowd surged forward around the men and they were knocked onto their backs, at which point my guys dove at them and started kicking them. Meanwhile, the policeman just kept wrestling with me and did nothing to stop the others. It was probably ten minutes before he restored order. I said, "I'll take these men to the police station. I want to make sure they are charged properly."

We tied the two men up and wrestled them up onto the back of the tractor-trailer, and I drove back to our main compound. But when I stopped, all of the women at Chimfunshi rushed out and attacked the men because they'd heard about what they'd done, and they dragged them off the tractor-trailer and began kicking and pummeling them all over again. Then, no sooner had I calmed down the women than Dave and our son Charles pulled up, and when I told them what happened, they too gave the guys a working-over.

Finally, we got to the house, and I was so angry that I told my guys to tie those two thugs to a tree and leave them outside for the bloody night. But Charles knew better, and said he'd drive them into town to the police station. They changed the wheel on the Land Rover and loaded the poachers into the back, and Charles was almost to town when he heard a commotion and turned around just in time to see a cooking pot come down on his head. The poachers had managed to untie themselves and were using the jack and other tools and anything else they could find for weapons. One managed to get away, but Charles caught the other one, tied him up again with electrical cable, and managed to make it to the police station. Then he drove himself to the hospital to have his head stitched up. The poacher got off with an awfully light sentence, I'm afraid, and Charles was left with a two-inch scar.

Another time, we were feeding the chimps when I looked up and saw a man and a woman running clumsily across the floodplain off in the distance, carrying what looked like elephant tusks. There were so few elephants left then that I became terribly worried, so I grabbed my shotgun and called my guys together and we raced out after them on foot. It was the dry season then and the ground was hard, but the dry grass was extremely high and it made for tough running. As we closed in, the couple veered toward the river and threw whatever they were carrying into the water. We noted the spot and proceeded to tackle the couple, and my guys tied them up; they were Zaireans. We then pulled two fresh elephant tusks from the Kafue, and it was clear they were trying to smuggle the cargo across the border.

I loaded the man and the woman into the large cage on the back of my truck and drove them to the police station in Chingola. There, a "kindly" police officer spoke to the woman, and, instead of arresting her, offered her a ride back to the Zaire border post! The man waited until he knew the woman had left, then said during interrogation that the tusks were not his but his wife's, and he got off, too. Frustrated and angry, I called a game ranger to come down and at least weigh, measure, and register the tusks, in the hope that something would go on the permanent record. But the tusks were stored briefly in the Exhibition Room, then reported missing altogether three days later.

The situation has changed because there are too many people now, too many poachers, and too many guns. When I think of the thousands of rhinos killed for their horns or elephants killed for their tusks, it makes me sick. There were once seventy elephants in this area alone. We used to drive from Chingola to Solwezi along a road that swings back and forth over a hundred miles, and we'd encounter herds of elephants all the time. In fact, we never knew how long the trip would take because we'd inevitably be forced to stop when the elephants bluffed and threatened us. It was as beautiful as it was thrilling. They'd migrate across the area every year, following the food as its ripened in each region, and they'd pass through our land as they worked their way west. To stand out on our verandah and watch the elephants just slowly grazing in the trees or drinking at the river was one of the most peaceful experiences I've ever known. But one year we realized that the herd seemed smaller, and then there were thirty left, then six, and today there are none. The last two elephants were killed a few years ago after supposedly raiding a local farm, but I doubt that's what really happened.

The forest came down, too. Where once these great forests stood, you can now see for miles, and the surrounding countryside has grown more arid and dusty as a result. Zambia was a lush, tropical country not so long ago, but you'd never know to look at it today. The Solwezi road is lined with charcoal bags waiting to be picked up by trucks that transport them for sale to the nearest town—bags that hold what's left of the trees. There are also dozens of logging trucks on the road each day, and people, always more people. Another disquieting fact is the number of villages that are springing up on the vast tracts of treeless ground. Even the so-called forest reserve areas are full of people who are constantly on the lookout for something to stave off starvation. Anything that flies, moves, or breathes is either edible or salable. When there is no food in your belly and no way to feed your family, nothing is sacred.

Bushmeat, the practice of hunting and eating apes, is totally out of control. I suppose it was inevitable that once the plains animals such as antelopes and bush bucks were killed off, mankind would eventually get around to wiping out what was left in the forests too, but the fact that chimpanzees and gorillas and monkeys are now being slaughtered is terrifying. Hunters regularly go into the forests with automatic weapons in search of chimpanzees, and they'll kill ten or more at a single go, chopping up the adults for meat. The babies—assuming they survive—wind up being sold as pets, but it's estimated that only one in five infants ever survives that ordeal, which means that if you count up the 80 or so chimps we have at Chimfunshi, there are another 320 that didn't make it. It staggers the mind. Because we live in Africa, we hear many blame the Africans, but it's not Africans who have brought about this disaster—it's Europeans. It's the European countries that offer debt relief for timber to countries in West and Central Africa, and European countries that are building roads into the rain forests. They are the ones who ultimately make a bad situation even worse.

Chimfunshi is sometimes referred to as a refugee camp for chimpanzees, but that's only because human beings have destroyed the chimps' natural habitat. And when we don't hunt and kill chimps directly, our internecine struggles do the job. The Democratic Republic of the Congo—the former Zaire—lies just eight miles north of Chimfunshi, and the constant wars and rebellions and violence of the last two decades have forced several chimp owners to surrender their "pets" to us, simply because they could no longer guarantee their own safety, let alone the chimps'. For instance, we were once asked by TRAFFIC, a South Africa–based animal welfare group, if we could accept three baby female chimps from Zaire. Apparently, a South African couple working in Zaire had bought these babies and cared for them until they went on holiday. But while in

South Africa, Zaire's civil war flared up again and the couple was unable to return. We did not know what we could do to help these babies, but we said we would try.

We had a visit around that time from two friends, Tristan Davenport and Vicky Borman. Tristan worked in Zaire, and when we told them our predicament with the Zairean chimps, they offered to help. Little did they know how involved they were about to get! The three babies were being kept in Likasi, a small town in the southwestern part of the country, and it just so happened that Tristan was planning to visit Likasi in the coming week, only a four-hour round trip from his home. We asked him to collect the chimps for us and carry them across the border, since someone with Zairean residency papers would have a better chance crossing the border with live animals than we would.

Tristan and Vicky had an awful time getting the chimps from their caretakers in Likasi, who kept demanding more and more money. But the babies were not in very good condition—they were suffering from severe malnutrition and a terrible skin disease of some sort—and Tristan feared they might die if he didn't get them out soon. After two weeks of intense haggling, threats, and frustration, a deal was struck: Tristan paid $600 for the three chimps, and I got a cable asking me to meet them at the Zairean border post of Kasumbalesa on the 20th of March.

The border post was a nightmare. We agreed to meet there at ten o'clock in the morning, but when I arrived, I could see it was going to be a very long day. The army had been ravaging the countryside and looting the villages, and everybody was fleeing. The rebels, meanwhile, were advancing on Lubumbashi, so refugees were streaming out of there as well. Government officials were lining their pockets one last time before they fled, and everything had its price, which was very high. If you did not have the money, then your goods—cameras, videos, tapes, or anything of value—were confiscated. Long queues snaked this way and that out of cramped government offices, and people were literally carrying all that they owned on their backs. Missionaries were walking by in tears. The soldiers had stripped their vehicles and taken everything, confiscating family photographs and videos and throwing them onto rubbish piles. Everybody seemed so desperate. Many people were selling their life's belongings just to raise bribe money, and being just five-foot-one, I kept getting lost in the crowds. Everywhere you looked there were soldiers—cruel-looking men in dirty fatigues, holding machine guns or pistols or machetes.

The whole scene was dark and sad, and it was very strange to be in the midst of so much human misery and be trying to stem chimp misery. But all I could think about that day was the chimps. It took Tristan and Vicky about seven hours to get through customs and immigration, arguing and haggling as they went. I was thrilled to see them at last. Unfortunately, the three chimps had broken out of their traveling cage because the soldiers at the border post had frightened them so, and their fear made them defecate all over the inside of the car. They'd also found a jar of chocolate and broke into that, so the car smelled like feces and chocolate. But brave Vicky, who was covered in excreta, was just so pleased to be out of Zaire with them that nothing else mattered. At one point, some soldiers walked up to the car and acted threatening, but Vicky had had enough.

"Fuck off!" she screamed, and the soldiers backed away.

Tristan said later, "You know, you could have gotten us both shot."

By the time we left the border post, Tristan and Vicky had been in charge of the chimps for nearly a week and had done a good job of feeding them and treating them for dehydration. But because of the problems and bloodshed in Zaire, Tristan and Vicky could not return there, so they settled in with us for a time at Chimfunshi. The babies, meanwhile, had clearly been through a terrible ordeal. Barbie was under a year old and the smallest, but she had no fur on her at all, and E.T. was so named because she had only one finger left on her right hand, which made her look a little like the alien in the Steven Spielberg movie. Roxy was the oldest and probably in better shape than the other two, but all of them suffered from scabies and other skin disorders, so we treated them with baths and ointments as soon as we got home.

Not all of my orphans have been so lucky, however. In 1990, a game ranger friend of ours named Bertie Roomer confiscated a small female chimp from some smugglers who were attempting to sell her in Kitwe. Bertie brought the chimp out to us, but it was clear from the start that something was not right. Although less than a year old, she was very small and undernourished, and she seemed to lack many of the natural responses to food and warmth and care that even the most mistreated chimps usually did. Her temperature also spiked alarmingly every few hours, then would suddenly return to normal. She could not be left alone, and would scream and scream if she went untouched for more than a few moments. We named the poor baby Sheena, and her sweet face and gentle disposition won my heart. But her condition had us all baffled, and I finally sent her to the University of Zambia Teaching

Hospital to see if they could help diagnose her trouble. Two different professors detected a heart murmur, probably due to severe anemia, but neither of them could determine the cause, and it was feared her heart would not recover from even the mild shock of anesthesia, so proper brain X rays were ruled out, too. We returned home and determined that we would simply do the best for her that we could.

We placed Sheena on a special diet that included lots of vitamins and antibiotics, and sometimes she seemed to be growing stronger. But the next day she would be down again. One day, it dawned on me that Sheena was not seeing properly, so I took her to an eye specialist, who did a very thorough examination. He told me that Sheena was blind. It seemed that both of her retinas had become detached, possibly due to a hard knock on the head she'd received during her capture. He also advised that an operation would not help. We left with a very bleak prognosis, but every time I tried to be stoic about it, I found myself overcome with emotion. To watch her turn and carefully step toward me at the sound of my voice—even though she was totally blind—was as great a display of faith as I had ever seen.

Despite her condition, Sheena was allowed to join The Babies, and really seemed to be at her best when out on the daily bush walks. She would sit with one of us on a log or on the ground while the others played, but would sometimes become adventurous and walk round and round unsteadily, always winding up in the same spot. She would pant with laughter if you tickled her, and sometimes just amused herself by playing with a leaf or a twig, rolling over onto her back and flailing her arms and legs in the air. And if one of The Babies actually approached Sheena and gave her a pat on the head or a quick tickle, she'd squeal with delight. We even let Sheena stay for extended periods in the cages with The Babies, who seemed to know something was wrong and were very careful with her. If Sheena climbed the mesh on the cage, for instance, Pippa would place her hand on her back and offer support. If Sheena became disoriented and screamed for help, Dora would rush over and throw her arms around her. At times like those, Sheena was as happy as she ever was, and she became so attached to Pippa that I eventually let her sleep with The Babies at night, confident that Pippa would keep an eye on her.

But Sheena's health was always fragile, and she developed both hepatitis and malaria less than a year after arriving at Chimfunshi. We removed her from The Babies' cage for health reasons, and I eventually took to carrying her with me everywhere on the farm. Then one day, I noticed a change in Sheena's demeanor. She did not seem to be getting weaker, but I think she was maybe ready to give up the very hard struggle she'd carried on for the past ten months. Her breathing turned erratic, and she became quite clingy, refusing to be put down for even a second. I was feeding her a bottle in the kitchen when she died, and she went peacefully. It was a very hard day, but a relief in so many ways. Sheena's suffering was finally over, but my hatred of poachers only deepened.

The cruelty of circus owners is another source of anger. We have received a half-dozen chimpanzees from wretched little circuses that can no longer care properly for their animals, and each comes with the sort of tics and phobias that only a lifetime of abuse can bring about. In late 1990, we received a telex from an American friend, Sarah Christiansen, who was visiting someone in Papua New Guinea and wrote to ask if we could take a pair of chimpanzees she'd found there. We were experiencing a shortage of cages at the time, but before she could receive our polite no, Sarah telexed again, writing, "It's too late. We're coming."

It seems a circus had gone bankrupt in Papua New Guinea at just about the time Sarah arrived, and her friend had asked if she wanted to go check on the animals. Sarah found Lucy there, a three-year-old female locked in a small cage without any food or water. No one knew how long the animals had been confined—some guessed as much as two weeks—but none of the animals had been fed for several days, at the very least. Sarah was apparently on her knees talking to Lucy and trying to comfort her when she heard a loud knock and a big hairy hand came over the back of the cage from another one behind it. Sarah walked round to that cage and found Toby, a male who looked to be about six years old, and knew then exactly what she had to do. But how she managed to wrangle export permits for the chimps and persuade the Shell Oil Company to pay for their transportation costs to Zambia remains a mystery to me.

Poor Toby looked so miserable when he arrived. There was hardly any fur on his arms or shoulders and he lacked any hint of muscle tone; even his ears failed to stand up properly. Locked for so long in a three-foot-high box, Toby could neither stand nor exercise, and his muscles had atrophied to almost nothing. We eventually refigured his age to be about ten based on his physical development, but he spent the first month with us hiding in a corner of his big indoor cage. Toby would scream loudly if any of the other chimps—even the little ones—looked his way, and he was seemingly terrified of everything. He refused to let his feet touch the grass, which was new to him, and it took him nearly six months to climb his first tree. But he eventually began to gain confidence, and then suddenly threw himself into the most bizarre physical regimen I have ever seen. As if trying to regain his circus physique, Toby began repeating all of his old routines, and would sometimes spend hours upside down on his shoulders, with his feet going round like mad, as though pedaling a bicycle. At other times he would lie on his shelf and flex his arms this way and that, or twirl his feet back and forth. And he always accompanied these workouts with a horrible noise—a sort of half-bark, half-growl—that we eventually came to recognize as his laughter.

We placed Toby and Lucy in a cage between the two young half-brothers, Choco and Leben, on one side, and The Babies on the other. Lucy spent most of her time snuggling up to Choco and Leben through the bars, and you could usually find one or both of them with their arms around her waist. But Toby, oddly enough, seemed to get the biggest kick out of The Babies. He was too rough with them at first, but a couple of nips here and there seemed to teach him that he couldn't have everything his own way. Toby was particularly fond of Dora, and loved grooming her through the bars, and we eventually learned that if we heard his bark-growl, it usually meant he and Dora were doting on one another.

Three years after arriving at Chimfunshi, Toby suddenly became very ill. There was no temperature, no swollen glands, nothing that I could identify, but he got quite lethargic and very picky about his food, and after about a month, his stomach became inflated and he passed no stool at all. We called a vet out to the farm and he anesthetized Toby, but after examining him, he advised that Toby was too sick to recover and should not be allowed to come round from the anesthetic.

Euthanizing Toby was one of the hardest decisions we ever had to cope with, but it was the right thing to do. A postmortem showed that he was suffering from enteritis, brought on by a perforated intestine, and his suffering must have been terrible. It was thought that at some stage in his life, Toby had been subjected to malnutrition and a heavy infestation of parasites and that his entire intestinal system was weakened as a result.

We went through a similar scenario in early 1994 when our old friend Karl Ammann, a wildlife activist from Kenya, contacted us and warned that an Egyptian circus was traveling through Africa and was expected to arrive in Zambia in March or April. It was thought that the circus was being used as a front for animal smuggling. It apparently gained entry into various countries as a "cultural exchange group" and therefore enjoyed diplomatic immunity, but was found to already have shipped eleven chimps out through Kenya in recent months. Chimps fetched as much as $10,000 apiece in Saudi Arabia at the time, and some of the circus's other animals, such as pythons and lions, were also prized as pets. The circus was duly escorted out of each country after much haggling with the Egyptian embassies, but not before a shipment of animals would be on its way to the Middle East.

Despite such a horrible record, the Egyptian circus arrived in Zambia in May, but could produce no documentation for its animals whatsoever. The Zambian government responded by confiscating them, and we were called to Lusaka to collect two chimps and an African gray parrot. The chimps, an eleven-year-old male named Boogie and a ten-year-old female named Tamtam, had since 1988 been kept in a tiny cage that measured just three and a half feet high by six feet wide by three feet deep. It was metal all around except for a small grid door they could look out of and a metal grid floor that allowed their droppings to fall through. Neither had been able to stand or move much for months, and, like Toby, their muscles had wasted away to almost nothing. Yet both had such distinctive faces. Boogie's was quite long with a thin muzzle, but the crown of his head was wide and his ears sat quite high and forward, almost like those of a bear. His eyes were also close together, giving him a worried, dopey look. Tamtam, meanwhile, also had a long, thin face, but her ears sat back like a human's, and she had very little hair on her face. They wore thick chains with padlocks around their necks, but as we had no keys for the locks, we decided to leave the chains on until the chimps knew us better.

We placed Boogie and Tamtam in a cage that must have seemed enormous compared with their old one, but they did not know what to do with the sacks and grass that we spread on the floor for bedding. Poor Boogie was so confused that he slept on the concrete floor for the first week, with Tamtam cuddled up alongside for warmth. We eventually spread even more grass on the floor, and the newcomers slowly learned to enjoy sleeping on it. Boogie also warmed to the idea of the sacks, and he enjoyed climbing inside and pulling it over his head, where he'd sit motionless for long periods, as if asleep. I used to stand and watch him for a while, then softly call his name. The sack would move a little, but you had to call him four or five times before Boogie would peek out and look your way.

Two months after Boogie and Tamtam arrived, a veterinarian visited us at Chimfunshi, so we decided to anesthetize the chimps and try and remove their chains. But when Boogie and Tamtam saw Dave with the dart gun, they went into hysterics, screaming at the top of their lungs and racing around looking for a place to hide. It was very obvious that they had been anesthetized before, or at least seen guns, and we felt terrible for putting them through such stress. But the chains were beginning to rust and we feared that Boogie and Tamtam might eventually develop bad sores, so we had no choice. We cut the chains off as quickly as possible and returned the chimps to their enclosure, but the matter was by no means closed. Both Boogie and Tamtam lost complete trust in us, and neither would look in our direction for days. After working so hard to get those chimps used to us, we had to start all over again from scratch, and I believe their faith in us was never quite as strong.

It was clear that Boogie and Tamtam's life in the circus had left more than physical scars. Tamtam came into oestrus regularly less than a year after joining us, but Boogie showed no interest whatsoever. I have heard that male and female chimps who are raised together from an early age regard one another as siblings and are unlikely to risk incest by mating, but Boogie's disinterest seemed to be more than that. He would ignore Tamtam completely when she was in season, even turning his back on her, and she would eventually approach us and beg to have her back scratched through the bars. It was as though Boogie's testosterone or sense of masculinity had been switched off.

Later, when we tried to integrate him and Tamtam with The Infants, Boogie was very unwilling to stake a claim for dominance. We opened the door into the enclosure, and waited and waited and waited. Finally, Boogie's head cautiously peered out, then disappeared. A while later, he looked out again. Eventually, Boogie came out slowly, looked about, and promptly went back in. Then he came out about a foot and sat down, just looking all around. I went to check on Tamtam and found her sitting inside the cage with her back to the door, as if she didn't want to look. She asked for a back scratch and I obliged, then tried to coax her out. It took a long time, but eventually she joined Boogie outside and, with arms wrapped tightly around one another, they went for a short "buddy walk" around the area. The Infants, meanwhile, regarded this behavior as ridiculous, and loped off happily into the trees.

Over time, Boogie and Tamtam began to adjust to life as chimpanzees. But both remained very low in the hierarchy, and poor Boogie was subservient to practically every other chimp in the group—including the females, even though they were half his age. Everything just seemed so difficult and confusing for him, and he finally died on the night of May 14, 1999. He'd been a bit mopey a few days before, but still drank his milk and ate some sweet potatoes. Then he became lethargic and I thought he might have the flu, so I gave him some medicine and an apple, and he went and lay down on a bed of straw and pulled his sack up over his shoulders. The next morning, Boogie was still lying there, and I thought he must be asleep. But when I approached to check on him, I could see he was dead. He had not been dead for very long and there did not seem to be any trauma in his face. I checked him all over for snakebites or any telltale signs of injury, but could find nothing.

Though I had no idea what'd caused his death, we chose not to have a postmortem done. I think maybe Boogie had just been so weakened—both mentally and physically—by his ordeals with the circus that even a hint of the flu was too much for him. Tamtam spent most of that day near the spot where I found Boogie and kept smelling the grass where he'd lain, though she ate well and in other ways seemed to accept his death without too much fuss. But I found Boogie's death to be terribly sad. He did not have a particularly happy life, I think, and I wish we could have done more for him.

Twelve

Birth

Liza Do Little with Ingrid

On the morning of January 10, 1991, Liza Do Little was missing, and we were concerned. She had skipped two straight meals—the five o'clock dinner the night before and then breakfast that day—and was nowhere to be seen when we stood and looked out from the viewing platform. The other chimps in The Wall were behaving oddly too, looking repeatedly out toward the trees as they ate, and we feared something must be wrong. Two visitors from Kenya, photographer John Richardson and wildlife activist Karl Ammann, offered to take one of our canoes and paddle downriver to see if they could find Liza that way, and they must have been gone close to an hour. At half past nine, Karl returned with a large smile on his face and extended his hand to Dave and me.

"Congratulations," he said. "You are grandparents."

Dave and I were so excited. Chimps don't always gain a tremendous amount of weight when they are pregnant—some barely show at all—and we had no idea that Liza was due. But we had thought she might be pregnant, since she had not come into oestrus for some time, and the books we had told us that chimps normally gestate for about 230 days. We were still learning as we went along in those days, however, and we hadn't begun to keep records as carefully as we do now. We scrambled into the canoe, but although we spent a long time on the river, neither of us could see Liza or her baby anywhere. In fact, she managed to keep herself hidden for the rest of the day. The next morning, Liza failed to come for her morning milk, and Dave and I were unable to spot her from the river. John and Karl said they caught another glimpse of her, but they could not see the baby, as she had it covered tightly with her arms.

Finally, around midday, I was feeding the chimps some mangoes when I looked up and saw Liza walking slowly up the path on three legs, with one arm clasped tight to her belly. She sat down at the mesh in front of the cages, and at first I could not see the baby and thought perhaps we had made a mistake, especially since Liza looked just the same as she always did. Then I realized that what I had thought was her stomach was in fact a shiny little baby whose coat blended so perfectly with her mother's that it was impossible to tell where one ended and the other began. Liza climbed onto the wire mesh in front of me, and suddenly this beautiful little face looked up and seemed to be staring right at me with smoky gray eyes. I am afraid I cried—the baby looked so perfect. Dave saw it later and was also moved, and he added that the hair on its head seemed so perfectly parted down the middle that Liza must have spent hours combing it for its first outing.

Unfortunately, Liza was not happy about our close scrutiny of her offspring. She took two mangoes and started moving away toward the trees, walking on three legs and using the fourth, as she always did, to carefully support the baby from underneath. She returned again for her five o'clock evening meal, but did not stay long. She seemed to get a bit agitated when the other chimps got too close, and we didn't get a clear enough look to determine the baby's sex.

On the third day, Liza came in at midday and got into a cage with Charley and Bella for lunch. Although a baby chimp's eyes are unable to focus at such an early age, this one certainly seemed to respond to noise, and consistently turned in whatever direction it was coming from. I told Liza how appealing I found her baby, who seemed to be looking right at me again. Liza was a lot more relaxed this time, and stayed a bit longer than before. (It must be noted that Charley, apart from patting her on the head in greeting, took no notice of the baby whatsoever—though he would soon become a great dad.) It was at this meal that the baby first made a sound—not a cry, but more like a contented mew—and we determined its sex for sure: It was a girl.

We did not want to name the baby until we knew its sex, and we had already agreed to name any offspring with the first letter of its mother's name. This was an idea pioneered by Jane Goodall in the wild at Gombe Stream; it makes record-keeping so much easier. For instance, Jane had named all of Flo's infants with "F"s, like Fifi and Figan and Faben, and it helps when you're trying to connect a certain mother to a certain baby. But this first infant had to be an exception. For all the work our dear friend Ingrid Regnell from Sweden had done on our—and the chimps'—behalf, we chose to name the first baby born at Chimfunshi "Ingrid."

Let me make it clear that we were not *trying* to get Liza pregnant. We have never forced our chimps to reproduce, and we do not place them together with breeding in mind. In fact, the word "breeding" is abhorrent to me in the context of chimps. Chimfunshi is not a chimp farm. We have never sought to make chimps reproduce in order to do anything with the babies, as if they were a crop. The orphanage is simply our humble attempt at giving chimpanzees back a little of what mankind has taken so brutally from them—land, food, family members, peace of mind—and our chimpanzees are free to choose their own partners and to procreate or not.

On the other hand, anyone who has studied them or read Jane Goodall's books knows that chimps, like humans, are extremely social creatures with very strong family bonds. In our experience, we have observed a fragmented group of chimpanzees come together and form a cohesive family group after the birth of just one baby. Reproduction somehow helps restore the social order in a chimp family, and that is why Dave and I have never interfered.

This places us far outside popular thinking in the animal welfare community. Most chimpanzee sanctuaries around the world insist upon using either birth control or sterilization, or strive to simply keep males and females apart. We know that our willingness to let our chimps reproduce has cost us grant money and some organized support, but Dave and I are doing what we believe is right. As long as we've got enough space so that the chimps don't feel overcrowded, and as long as they have the right environment in which to have babies, well, why not? Our whole idea was to try and give them a normal life. And to me, that normal life includes having babies. In fact, if I don't allow them to have babies, I can't really see the point in our efforts; it'd feel like they're all just sitting here waiting to die. Too, even though they say birth control implants are fairly effective, it doesn't take into account all those males with all that testosterone running around. And if we simply go ahead and castrate the males, we don't have chimps anymore—we've created something else altogether, something less.

I sympathize with those sanctuaries that do have space limitations and are forced to use birth control. If somebody's taking in chimps in order to give them sanctuary, to save a life, and they haven't got space to grow, I can understand that they might not allow the chimps to reproduce. But we *do* have the space. I've got thousands of acres out there that I can use any way I want in order to help chimps, and as long as I can get the money to build more enclosures we'll keep on like that. I'm convinced that our chimps choose to reproduce only because they know it fits in with the balance of nature. I believe most animals will regulate their birthrate in accordance with their natural surroundings. If there's not enough food or space, there is a higher rate of death among infants, or some are born weak and soon die—there are a number of things that occur that keep the numbers reasonable. I'm afraid it's only we humans who choose to ignore the natural laws.

Baby Ingrid's arrival had been a surprise—and a wonderful one at that—but we really didn't know what to do afterward except sit back and watch Liza learn to be a mother. She was quite secretive for the first ten days, covering the baby with her arms or legs when we took too much notice of her, but we did hear the baby make a soft "*Ooh, ooh, ooh*" sound on several occasions. Eventually, we saw Ingrid climb up Liza's tummy searching for a nipple, then suckle for a good three minutes before letting go. As she did this, I did my own form of "oohing" and "aahing," all the time telling Liza what a wonderful baby she had. Suddenly, she turned to look me straight in the eye, then pushed herself closer to the bars on the cage window until the baby was in full view. I carefully put my hand out, and Liza watched as I stroked the baby on the arm two or three times.

The chimps were just as curious as I. I watched Charley try to touch the baby on the head while he and Liza were sitting together eating their lunch. Liza let him touch the first time, but the second time, she very firmly moved his hand away. When Liza's attention was elsewhere, however, Charley actually put his big fat finger into the baby's mouth, quite gently, then withdrew it and patted the baby on the head. Once again, Liza firmly moved his hand out of the way.

When Ingrid turned one month old, we got an unexpected break in the weather. January and February are among our rainiest months in Zambia, and it seemed as if we'd gone weeks without seeing the sun, leaving everything damp and musty. But the rains finally let up one day and the sun shone through and all the chimps seemed to celebrate. In fact, they were all late for lunch because they got so totally involved playing games around some bushes and a huge puddle. Charley was lying on his back in the long grass near a bush with about six chimps running round and round the bush past him, and he was laughing and trying to catch them as they zoomed past. Tara got caught twice, rolled around with Charley, then got back into the game of running around the bush. Donna seemed to catch hold of Charley's hands a few times, do a bit of a jump, then be off running again. Rita was not so brave. She seemed to hit Charley's foot in passing, but never got near enough to his hands to be caught.

During all this, Liza was sitting on a low branch nearby, just watching and nursing Ingrid. But I think watching all those chimps enjoying themselves got to be too much for her. She climbed down the tree and got in the way of Tara as he rushed past on his way to Charley. She grabbed hold of him, and the two of them started rolling about in the grass, both laughing loudly enough for me to hear them. I was a bit concerned about the baby, but Liza was not the least bit worried. I could not see the baby at all, only a glimpse now and then of its little fingers holding on to Liza's fur. Both Liza and Tara were rolling in and out of the big puddle, and the baby must have gotten awfully wet. This went on for a good five minutes before Liza decided she had had enough and came to the cages for some food. Her fur was wet, and so was Ingrid's, but they both seemed perfectly happy.

The birth of Ingrid seemed to touch off a sort of baby boom at Chimfunshi, which can probably be attributed to a number of factors. First, our oldest chimps were all becoming sexually mature at about the same time, meaning that not only Charley and Liza, but also Girly, Bella, and several of the males were suddenly of age. Second, the move to the seven-acre enclosure had given the chimps the space and security to begin to reproduce, if they so desired, and Charley had responded by asserting his dominance and consorting with whichever females were in oestrus. Third, the social structure was beginning to fall into place, not only for mothers and fathers, but also for the younger chimps, who could serve as aunts and uncles and learn valuable parenting skills from their elders.

Girly was the second chimp to give birth at Chimfunshi. We knew she was pregnant, and when she failed to come in for her evening meal on September 20, 1991, we figured things were imminent. Later the next day, as we were busy handing out food, Patrick said, "Look at Girly walking toward the cages, as though she's holding her stomach." As she got nearer, we were able to just make out the top of a tiny little head. She held it so close that we could not see anything else of the baby at all.

Ingrid Regnell arrived from Sweden for a visit on September 22, the day after Girly gave birth. It took us nearly two weeks before we could say for sure what sex the baby was, but it was Ingrid who first saw a tiny little penis and excitedly remarked that it was a "Goliath." That remark left us all wondering about Swedish men—as well as prompting us to call the baby Goliath, even though he was so tiny. He seemed to cry a lot the first few days, but as Girly's confidence as a mother increased, the baby's crying subsided. He had an exquisite little face, with lovely blue eyes and a habit of bobbing his head up and down, as if he were agreeing with everything you said.

Goliath nearly came to a tragic end before he was old enough to stand. One day I went over to The Wall to give the chimps their lunch and, after being greeted by the usual shouts and welcoming food grunts, had entered the center aisle between the cages carrying a boxful of fruit when the noises suddenly changed from happy to angry. Something had gone wrong in Charley's cage, and when I looked inside, I could see that he was jumping up and down on a chimp on the floor. Whoever was on the floor was screaming terribly, but managed to get out from under Charley and move away to the other side of the cage, and that's when I realized it was Girly. Somehow, Charley had got hold of Goliath, and his huge hand was squeezing the baby tightly around the chest. But instead of continuing to battle Girly, Charley turned and fled through the open door like a shot, out toward the trees in the enclosure, holding the baby high above his head. He did not get very far. Girly was out of the cage as fast as Charley, the whole time screaming at the top of her voice, and Spencer, Pal, and Liza must have somehow known what was happening, because they all came running out from different directions, too.

Charley was running with the baby high in the air, and all I could really see were Goliath's little head and tiny little hands and legs just flopping up and down; the rest of its body was covered by Charley's enormous hand. I felt so sick that I could make no sound, and I felt sure Charley must have injured the baby already—I was certain that its tiny ribs must have been crushed by his strong grip. All the chimps were making a terrible deafening noise, except for the four who were hot on Charley's trail. Pal got in front of Charley, which made him change direction, but then he came face-to-face with Liza. He turned again, and there was Spencer. Then he turned and went off around a bush, but Girly was faster than he was. I was sorry that the bush obscured my view; I did not see how she did it, but somehow or another, Girly got her baby back.

Girly ran toward the bushes away from Charley, but stopped when Pal ran toward her. I was amazed how solicitously he patted her shoulder, as if to calm her nerves; then he leaned right down to the baby and took hold of its arm, Girly doing nothing to stop him. He seemed to have a good look, then made a funny noise and walked away toward the feeding area. In the meantime, Charley had come back to his cage, with Liza following him. I felt very shaken, but could do nothing except feed the chimps as if things were normal. After about ten minutes, Girly came into the cage, cautiously greeted Charley, and accepted some food from me. The baby seemed none the worse for the traumatic few minutes he had spent with Charley, but he did fall sound asleep during lunch, which was unusual.

Despite that outburst, Charley was usually an amazingly gentle and attentive father, especially considering that he was probably taken from his family at too young an age to have learned much about parenting. I once saw Liza go up to Charley and greet him by presenting her full forearm to Charley's wide-open mouth. He mouthed it gently a few times, and Liza withdrew, but not before Ingrid had pushed her forearm at Charley's mouth, too. It made me shudder when I saw his huge teeth close and this tiny little arm disappear in his cavernous mouth. But he was making lovely play-type noises and opened his mouth quickly when the baby got a bit anxious and tried to pull her arm away.

The fact that Ingrid was the first baby at Chimfunshi—and, for a time, the *only* baby at Chimfunshi—meant that she was probably coddled a bit too much by the others. For instance, most of the chimp books I'd read stated that a baby usually starts to travel on its mother's back jockey style at about four months of age. But Ingrid was clearly the exception. She was almost a year old before I saw her riding on her mother's back, and she only did it then because Liza refused to carry her underneath her stomach anymore. Interestingly, the very next day I saw Goliath riding on his mother's back in exactly the same way. I can't say for sure if it was mimicry, but I do know that watching the two mothers walking about on the paths with their babies on their backs made for a lovely picture.

When a baby is born at Chimfunshi, we go out of our way not to handle it. If we need to take the baby away from its mother to save its life, we sometimes do, but I don't want these second-generation chimps becoming too familiar with humans. Their mothers usually know how to care for them and give them what they need, so let them, I say. It's my hope that by the time we reach our third generation—maybe in another five or ten years—they'll be back to something like wild chimps again, chimps that have no need or desire to come in contact with humans. But in the early days I spent hours playing with Ingrid and Goliath through the cage bars, working to earn their trust, simply because it made me so happy to see the lives of these chimps coming full circle.

I used to feed Goliath and Ingrid with little pieces of oranges or bananas (I believe that their mothers were indulging me a bit in allowing me to do so). At first, Goliath was very hesitant about taking any bits from me, but soon he learned to make lovely little food grunts whenever he saw me coming with a banana. He and Ingrid also became very adept at climbing the bars of their cage at feeding time. While Girly and Liza sat and ate, the babies often climbed above their heads and played with each other, both of them swinging about and holding on with one hand. They would catch each other round the neck or grab the other's arms or legs, all the while wearing play-faces and making little laughter sounds. The mothers seemed to ignore them until something threatened—such as Charley rushing in—whereupon Liza and Girly each grabbed her baby and prepared to run away if necessary. One day, Charley came crashing into the cage all worked up, and the two mothers got ahold of the wrong babies; it was rather funny watching the babies squirming to get back to their proper mother. Liza had just gotten through the door when she realized something was wrong, so, forgetting all about Charley, she rushed back into the cage, virtually threw Goliath at Girly, grabbed Ingrid, and was off.

Of all The Youngsters, Cora was the most interested in the new babies. She pestered Liza for weeks in an attempt to touch Ingrid; then, once Goliath was born, she switched her focus to him. But whereas Liza was cautious and unwilling to let Cora handle Ingrid too much, Girly was content to let Cora baby-sit whenever she wanted. Before long, Cora was walking around with Goliath clinging to her back, and once she sat and played with Goliath next to the cages for a good ten minutes before his mother made an appearance, at which point Girly just sat down next to Cora and watched for a while before picking Goliath up and walking away. We even saw Charley walk up a path once with Goliath holding on under his tummy, though Girly was not far behind. She clearly took a laid-back approach to mothering.

Big Jane was the next chimp to give birth—our third baby in just under a year and a half. On July 29, 1992, Big Jane decided that she was going to spend the night in one of the cages, which was most unusual since she was always one of the first to go out and settle in the trees for the night. We decided to watch her carefully, but in the morning she had her milk and behaved quite normally. The next night she again opted to sleep inside, but on the morning of the 30th when I went to bring her some milk, she was nowhere to be seen. All the chimps came and had their milk, including Spencer, who finished his cup and then went into Big Jane's cage and began sorting through the straw where she made her bed. The next thing I saw was him lifting something up carefully and moving quickly toward the door. As he passed by, some of the other chimps became curious about what he was carrying, and it was then that I realized he was holding a placenta. He held it by the cord and it seemed to be about eighteen inches long.

We knew then that Big Jane must have given birth, but we did not see her until the evening meal, when she sneaked in, carefully clutching something very small against her stomach. She seemed frightened by the attention the others were paying to her, so we decided it would be a good thing to keep her in for a few days until we knew for sure that she was able to feed and cope with an infant. We very soon saw that it was a female, and a name had to be chosen that had the same initials as the mother. We thought of all sorts of things, but none seemed to fit until a visitor thought of "Bijoux," which is French for "jewel."

Big Jane was locked in her cage for three days. During that time we saw the baby suckle quite a number of times, and we could also see that it had a very strong grip when Big Jane moved quickly and it had to hold on. We let Jane out on the fourth day, and she got a tremendous amount of attention from all the others, but seemed to have gotten over her initial nervousness. It was funny—the big males seemed more interested than the females, and kept going up to Big Jane and trying to gently touch Bijoux. She allowed some, such as Pal, Spencer, and Tara, to have a good look and a little pat, but others she pushed away. Big Jane would not move without first wrapping her left arm tightly around Bijoux, and she sometimes held her so close that it was impossible to see the baby at all. She was obviously a good mother.

It was not long before Ingrid, Goliath, and Bijoux had become the best of friends. They'd spend hours playing together in a little ball of black fur and pink toes, rolling about in the straw and laughing, and the three mothers would sit nearby, watching contentedly. But trouble would invariably break out at mealtimes, when I offered the babies their special treats. Ingrid always assumed she should have hers first, either because she was the oldest or because she just felt dominant. If Goliath got a piece of something that she wanted, she'd scream and then bang him on the back. Goliath usually responded by shouting and hitting right back, and before long a fight would break out. The noise could be deafening from those two little ones as they'd swing wildly at one another, connecting on every third or fourth punch. But the battle only lasted a moment or two, after which they'd break, spend a second panting heavily and glaring at one another, and then rush back together in a loving cuddle. During these gyrations, Bijoux would retreat to sit between her mother's legs, watching with wide-open eyes as if shocked by the proceedings.

Not every birth at Chimfunshi went so well, however. About two months after Big Jane's delivery, I went early one morning to check on the chimps and found Tobar playing with what looked like a placenta. I had suspected that Bella might be pregnant, and assumed that she'd delivered sometime during the night. We could not find Bella anywhere, but when Girly came in for milk, she pushed this horrible-looking mess through the bars. It was the remains of a baby, but there was very little left of it. None of the chimps seemed very excited about it—all was eerily quiet. Bella never did come in for breakfast, but she finally showed up around lunchtime. She ate very little food, and looked thin and wan. It had obviously been her baby, but it was impossible to say why the baby had died. Bella remained sickly for a few days, until I gave her some medicine and vitamins.

Rita was another chimp who had tremendous difficulty with her initial pregnancies. Considering her excellent maternal instincts and penchant for mothering even chimps that were older than she, I was excited when it became clear she was pregnant. But as I was preparing to feed the chimps one morning, I saw Rita carrying a dead baby in her hands. She looked very sad and seemed somewhat pale, but by the time I got the food ready for breakfast, she was sitting alone in a tree without it. Then I realized that Girly had it, and it was quite horrific to watch her sort of half-playing with this brown, bloody mess. Finally, she came into the cage and pushed the dead baby under the bars of the cage and made a flicking motion with her hand, as if telling me to take it away. We will never know what caused the premature birth, but Rita eventually recovered her strength and got back to normal.

About seven months later, I was awakened by the chimps' wild shrieks and cries, the sort that indicate something is happening that they don't understand. It was just before dawn, so I got out of bed and went over to the seven-acre enclosure to see what the matter was. It was still fairly dark but the chimps knew I was there. Rita came close to the window, and I could see she had something in her hands. At first I thought she must have killed a bush baby, and it was not until later, as the light came up, that she pushed this dead body toward me and I realized it was another dead chimp. Rita had been about seven months pregnant when this miscarriage occurred, and, as with the previous one, we had no idea what caused it. She was depressed for a time and would not eat much, and it was clear the loss of a second baby had affected her a great deal.

It's very sad when an infant dies, but we try not to interfere too much. It is part of our belief that these chimps should lead a normal life. Humans have babies who die, and so do our chimps. It's a fact of life, and we try to respect that process. Most of our chimps who have had a baby that died carried the body around for two or three days afterward, as if to nurse it. People ask why we don't take the body away and do a postmortem. My firm conviction is that if this is their way of mourning, their way of getting over the loss, then so be it. And if the baby miscarried for some reason, though we really and truly don't know why it occurred, we do believe it was supposed to be that way.

We've only intervened when absolutely necessary, such as the time Noel gave birth about four years after arriving at Chimfunshi. Noel was approximately eighteen years of age at that time, making her the oldest chimp to become a mother at the orphanage, but the fact that she'd been so gentle and loving with The Babies made us sure she'd be an excellent parent. We were terribly wrong. I happened to be away in South Africa at the time, but Dave said it was horrifying to watch her throwing her baby carelessly around her neck, holding it upside down, and generally treating it like a rag doll. In fact, she treated the poor thing so wretchedly that Dave decided the only answer was to take the baby away and rear it by hand. Dave gave Noel an anesthetic, but she fell on top the baby as she went down and Dave became concerned that she might smother it, so he went into the enclosure and tried to take the baby away before Noel was completely out. The baby screamed and Noel sprang to life, tackling Dave as he ran for the door and biting him severely several times on the leg. Some friends managed to pull Noel off Dave and drag him through the cage door, but by this time there was blood everywhere, and he had to be taken to a hospital to have his leg stitched up. He still looked terrible when I returned a week later.

A neighbor did her best to get the baby sorted out, but the poor little thing died during the night. Noel finally came around from the anesthetic and seemed none the worse off; nor did she seem to miss her baby. I was really surprised by her behavior, since she was the one who'd so readily adopted The

Babies, looking after them when anyone got rough, playing with them, and sharing food with them. She also took a keen interest in the other babies born at Chimfunshi, and one would have thought she'd learned something from seeing them raised. But perhaps there was something not quite right with the baby and Noel knew something we didn't.

Tina was another chimp that just would not—or could not—get the hang of being a mother. In 1992, she and her mate, Charles, were living at the Mundawanga Zoo in Lusaka, a horrible, outdated facility that lacked both financial resources and staff. The zoo had experienced several breakouts, so they asked if we'd take Charles and Tina on a temporary basis while they searched for funding to upgrade their cages. Normally, Dave and I would never agree to "boarding" chimps and we have no desire to loan them out, but we honestly thought we could strike a deal with the zoo once the chimps were at Chimfunshi and that would be that. But it was not to be.

Despite our best efforts—including offers by our supporters to purchase the chimps outright—the zoo insisted on reclaiming Charles and Tina after just a few months. What a dreadful day that was, for those poor chimps to have been given a glimpse of how life might be, and then have it snatched away. As that truck drove away with Charles and Tina in crates, Dave and I vowed right then and there to never accept a chimp unless it was on a permanent basis.

Once back at the Mundawanga Zoo, Charles and Tina suffered horribly. Their cages may have been reinforced, but the country was in a deep depression at the time and the zoo was practically abandoned. There were stories in the press about lions eating one another to survive and about other animals not being fed for weeks at a time. Charles and Tina survived only through the generosity of some local volunteers who fed the animals at the zoo out of their own pockets for nearly eighteen months. After much haggling and threatening, we finally managed to secure Charles and Tina's retransfer in early 1994.

But even though Tina soon regained her health and fitness, one thing she could not recover were her maternal instincts. She'd lost one baby due to zoo mismanagement while at Mundawanga, but she arrived the second time at Chimfunshi accompanied by Tembo, her two-month-old son. The baby seemed healthy and happy, and we looked forward to his joining our nursery. But Tina had other ideas. She was a wretched mother, either throwing Tembo all over the cage or leaving him alone and neglected in the middle of the floor. We also saw Tina on many occasions pushing the baby away from her nipples and not allowing it to drink. We could see that the baby was becoming very dehydrated and weak, but we were struck with a flu epidemic at that time and found ourselves working around the clock just to keep some of the other adult chimps alive. By the time we finally anesthetized Tina and took Tembo away, the baby was clearly near death. For two days I carried Tembo everywhere and fed him milk and baby food laced with vitamins, and, for a time, he seemed to respond. But just after midnight one night he suddenly sat up in bed, screamed, and died.

It wasn't long before Tina was pregnant again, and she gave birth to another baby boy. We called that one Thompson, after one of our supporters, Steve Thompson, but Tina repeated the same mistakes with him as she had with Tembo. When she wanted attention, Tina would scream and shout and beat her son mercilessly. She also flung him around the cage, and hit him frequently on the head; the hair still won't grow back over one scalp cut she inflicted. We stepped in and removed Thompson for his own good, even though he was only about five months old at the time. But he weighed no more than five pounds and had clearly been through a lot. He was desperate for anybody to comfort him, and we raised him in our house until he was old enough to join the group we called The Little Ones in the nursery.

By 1997, we'd had fourteen babies born at Chimfunshi, including Liza's second, Lori, who arrived in 1995, and both the chimps in the seven-acre walled enclosure and those in the fourteen-acre enclosure were coming together as stronger family units as a result. But nothing gave me more pleasure than the morning of January 10, 1997, when Patrick came running to the house yelling that Rita had had a baby. My first thought was that she must have kidnapped someone else's infant, but after examining her closely we ascertained that it was definitely her own. After the failure of her second pregnancy, she'd come into oestrus again almost immediately and must have conceived then. Thus, exactly six years to the day after Ingrid became Chimfunshi's first baby, Rita produced number fifteen. It was a healthy little female that we named Renate, after another of our big supporters, Renate Winch, and Rita clearly knew what to do. She fussed and cooed over that baby for hours that first day, and always seemed anxious to show Renate off.

And if I harbored any doubts that Rita might not raise Renate in the proper style, they were dispelled one night around 7:30 P.M. when I went into our bedroom and found Rita and Renate sitting in the middle of the floor, with most of the things off my dressing table in front of them. They had opened all my jars of cream and tins of powder and makeup and tested each for "edibility," then had opened a bottle of perfume and spilled it all over the place, dousing each other in the process—both chimps smelled strongly of hyacinth for the next three days. The chimps had also found a cache of Dave's chocolate bars and eaten them messily on the bed. But Rita was not the least upset about me catching her in the room. Instead, she held out her hand and greeted me happily, then took Renate by the arm and walked casually past me and out the door. Clearly, Renate's education was just beginning.

Thirteen

The Attack

Choco (left) with Leben

Mondays are always difficult at Chimfunshi. That is the day the locals come to sell us their produce or fruit from the forest, and the crowds begin arriving early. We have asked people not to come before seven o'clock, but it does little good; some days the line starts forming at the main gate as early as six, with everybody wanting to be first. There's always lots of excitement and chatter and nervousness, especially if Billy the hippo has been up and about, and we've been forced to bring our staff in early on those days just to keep order.

One recent Monday, a crowd was already queuing up before dawn, despite extremely heavy rains that had fallen for weeks. Unbeknownst to us, a group of about fifteen Congolese had used a boat to ferry illegally across the Kafue River, putting ashore right alongside the seven-acre walled enclosure, intending to sell the fruit they'd brought over. I heard them talking and laughing as they walked past our house along The Wall, but I didn't give it much thought. We'd had plenty of Congolese on our property before, and even though they were technically breaking the law and in the country as illegal aliens, there really seemed to be no way to stop them.

Anyway, I was more concerned about Dave—he'd been hit with a particularly serious case of malaria and had spent much of the last week in and out of the hospital. He was racked by cold sweats and a rapid pulse, and his sleep the last few nights had been fitful at best.

I was busy feeding the parrots when, all of sudden, the main compound exploded in a panic of terrible screams and cries and people seemed to be running every which way. The chimps were shrieking, too. There were bags of fruit spilled all over the place, and everybody seemed to be shouting in languages I did not understand. It was clear something had gone wrong. Then one of our keepers, Dominic Chinyama, came running up to me and shouted, "Come quickly! Leben has killed a woman!"

My heart sank. Leben, who arrived at Chimfunshi along with his half-brother, Choco, from the Tel Aviv Zoo in 1990, had always been an excitable chimp but hardly what you'd consider aggressive, and he'd never bitten or hurt anyone. In fact, he'd always been something of a "softie," in my opinion, too habituated by captivity and hand-rearing to ever really compete with the wild-caught chimps at Chimfunshi. When he and Choco first arrived, they were actually so sullen and unresponsive that I feared they might be a little retarded, and they clung to each other like magnets. Only when a visitor from Israel spoke to them in Hebrew several months later did they spring to life and and begin hugging one another and laughing out loud. It turned out they spoke Hebrew, not English, and they always remained a little different from the rest of the chimps at Chimfunshi.

I did not see the attack, but as I raced around the house toward the vendors, its awful aftermath was clear. The Congolese had apparently teased the chimps and thrown rocks at them when they passed the area in The Wall where it's possible to see inside, because when a small portion of the wall suddenly gave way—caused, no doubt, by the rains that had left the foundation weakened—the chimps inside emerged terribly angry and excited and confused. Most of the people screamed, dropped their bags, and fled. But one unfortunate woman stood her ground and was indeed attacked by Leben, who went mad when she tried to fight him off. He made an awful mess of her, ripping off her scalp and tearing away a part of her face, then biting two fingers off her right hand and the big toe off her left foot. Leben dug into her stomach and back with his teeth and nails, opening up large gashes that were bleeding profusely, then dragged her away crazily, like a rag doll by the wrist, back toward the cages. He stopped long enough to have another go at her, tearing away her clothes, and only retreated a bit when a mob of screaming people began to close in and throw objects at him.

Grumps, who also escaped, ran up suddenly and tried to pull Leben away, but Leben shrugged him aside and stood over the woman's body. It seemed certain that she was dead. Grumps persisted, however, pulling and tugging and leading Leben back toward the enclosure. Leben's face was split in a grimace of fear, and it was clear he was not himself; indeed, he looked like no chimp I had ever seen.

I screamed for Dave, telling him to get the shotgun in case it was the only way we could stop Leben, and I do believe that at that moment I would have allowed Leben to be killed. Our chimps had never taken a human life before, yet here it was happening right before my eyes, and I didn't know what else to do. Leben had crossed the line; this wasn't like injuring Dave or myself, this was a stranger, one who was not at fault. Leben had suddenly placed everything we'd worked for at Chimfunshi in the balance, and all I wanted to do was end the incident before it got any worse—even if that meant taking the chimp's life myself.

Oddly, some of our staff was extremely upset by my order, including Dominic, who was practically in tears as he ran in front of me, trying to hold me back.

"No, Madame, you cannot kill Leben!" he cried. "This is a sanctuary! This is his home!"

Scrambling for ideas, I decided to try and scare Leben away by driving at him with our largest truck. So I climbed up behind the wheel and drove toward the enclosure, where I was able to chase Leben by honking the horn and driving as close to him as I dared. Suddenly, the poor woman moved slightly, and we realized that she was still alive. My mind switched gears in an instant, and all that mattered was saving her life. One of our keepers, Ryan Mumba, picked the woman up and put her in the front of the truck, and with him holding her, I drove back to the house. We carried her into the bedroom and laid her out on the bed.

I radioed for our head keeper, Ian Visser, to come down to the main compound as quickly as possible, since he had taken a paramedic first aid course and might be able to help. But it was clear we needed to get this woman to a hospital at once, so we made a bed of sorts out of blankets and mattresses in the back of Ian's pickup truck and set off immediately for Chingola at high speed. Dave radioed to town to ask them to alert the hospital, and I rode in the back along with Edwin Madichi, a member of our staff who spoke some of the same Swahili as this woman. Amazingly, she spoke coherently on the ride, if in a very low whisper, and asked Edwin to take down some notes for her, as sort of a last will and testament, I suppose. She asked that he please tell her husband's family that she came across the river to this farm of her own free will and that what had happened had nothing to do with him, so he should not be blamed. She also said that in the pocket of her shirt was 6,000 kwacha, which she asked be given to her husband. She said that her name was Shanta Chipalo.

Someone had managed to contact her husband, and he met us in Chingola. We got to the hospital, where the doctors took her straight into the operating room. We were told that her blood pressure was almost nonexistent and that the chances for her survival did not look good. Ian and I had an anxious five-hour vigil in the halls outside the operating room, while three doctors did what they could. She eventually emerged swathed in bandages—but in stable condition. The doctors were exhausted, but said that she might live, provided infections were kept in check.

Meanwhile, back at Chimfunshi, Dave; my daughter Lorraine; her husband and our foreman, Ian Forbes; and the rest of the staff mixed an anesthetic into some drinks in plastic bottles in order to tempt the escaped chimps, hoping that once drugged they would be easier to handle. Leben was clearly the biggest problem. He ran off away from the house during the commotion, though it was thought he might have grabbed a bottle as he fled. We'd never had a chimp attack before at Chimfunshi, and nobody knew what it meant. Would Leben attack again? How long would it be before he calmed down? Could he be trusted out in the open? Worse, what if somebody else got hurt? Luckily, most of the chimps were recaptured without too much fuss, but Leben remained on the loose, only to be glimpsed dashing here and there through the heavy green undergrowth throughout the day.

Finally, Leben was spotted late in the afternoon near the food storage hut at the seven-acre enclosure, where he ripped the locked metal door from its frame. The hut had nothing inside, so he lost interest and headed toward our house, but it quickly became apparent he had indeed drunk the anesthetic and that the drugs were beginning to take effect. Dave, although severely weakened and disoriented by the malaria, managed to shoot Leben with an anesthetic dart, and they gathered the chimp's inert body and placed him inside a traveling cage in order to return him to the enclosure. But before they could close the cage door, Leben suddenly sprang to life again, wriggled free, and started running around madly. He was not right and kept running into trees, and slammed into one so violently that it knocked him down long enough for Ian to dart him again. They wrestled Leben back into his cage around 6 P.M., just as Ian Visser and I returned from Chingola, and we all struggled to take in the day's events.

I am a firm believer that you never give up, and that even the worst situation can be dealt with and put behind you. But I was devastated by Leben's attack, and found myself crying uncontrollably later that evening. It was probably as sad and low as I've ever felt at Chimfunshi. I kept asking myself why this had happened. Of all the escapes over the years, nobody had ever gotten seriously hurt, and usually Dave or I could step in and physically lead a chimp away from trouble, even if it meant placing ourselves in considerable danger. And why Leben? After all those years of Charley and Chiquito and the other big males getting out, why would Leben be the one to hurt somebody? The attack underscored the wild, unpredictable nature of chimps—including those born and raised in captivity—but I still found myself thinking that if I'd been there a few moments earlier, Leben might have listened to me and stopped, or at least done less damage, or even wishing that Leben had attacked me instead. Somehow, I thought, that would have been easier for me to understand.

I was also terrified at the prospect of what this would mean for Chimfunshi. Surely, word would get out. Would the police come? Would they take Leben away? Would they kill him? Would they try to shut us down? Would they arrest us? My mind raced crazily through all sorts of terrible scenarios that night, each one worse than the last. I was on the edge of despair—I was sure we'd lose the farm. I could just see all of our chimps being shipped off to laboratories and zoos, or worse, being destroyed as a public menace; I felt all our years of hard work and devotion slipping away. I do believe my heart was breaking as the fear and sadness of the day overwhelmed me.

But one thing a lifetime in Africa has taught me is that despair is of little use. This continent contains more beauty and joy than any other, I told myself, though it often comes at a terrible price. And that's when I realized I would have to dig deep within myself and find the strength to deal with whatever arose, and somehow that gave me peace.

Taking care of Shanta became my top priority. She was placed in a special care unit at the Chingola hospital, and I went to see her as often as possible. The doctors were amazed at her resilience. She had long conversations with the nurses and told them all about her life and family back in Lubumbashi, and moved about the room and ate well. Shanta did not complain very much, even though having her bandages changed every day must have been a terrible ordeal. The doctors began skin grafts to repair some of the damage, once her wounds healed, and she needed plastic surgery to fix the structural damage done to her face. Of course, we had no insurance for an accident of this nature, but we covered her medical costs and did what we could to support her through the ordeal. I brought the doctors fresh bandages and gloves and medicine that they might otherwise not be able to get, and Shanta always received me warmly when I visited. God, how I admired her strength and courage.

Although Shanta was in Zambia illegally at the time of the attack, the local police nevertheless opened a file on her case, and monitored her recovery—and our support—closely. But as long as we did all we could to help her get better, they seemed satisfied.

Leben, meanwhile, returned to his sweet, docile self almost immediately and showed no aftereffects from the attack. But one thing was clear: The Great Wall of Zambia, which had come to symbolize our commitment to chimps so early on, was too old and fragile to be effective any longer and had to be knocked down. At Chimfunshi, changes needed to be made.

Fourteen

The Transfer

I was a bundle of nerves as we prepared the chimps for transfer to the new enclosures.

Even though our first chimps lived in large cages alongside our house, we always dreamed of giving them more space and allowing them to behave as much like chimps in the wild as possible. Dave and I were blessed with plenty of land on the farm, and it became obvious over time that large, fenced enclosures were the only real solution. The seven-acre enclosure came first, and when we put nineteen chimps in there in 1989, Dave and I breathed a big sigh of relief and said, "There. We've done it." And when we added the fourteen-acre enclosure in 1991, Dave and I breathed another sigh of relief and said, "There. We've done it again." The five-acre enclosure followed in 1997, but by then, neither Dave nor I were sighing in relief. We knew we weren't finished yet.

All along, Dave and I kept discussing building something bigger, something that would really allow the chimps to come and go as they pleased. For even though they had plenty of space and trees to climb in the enclosures, most of the chimps were still too reliant on humans. They were fed daily and, because most of them had known us since they were infants, they came to look for us every day, both as friends and protectors. Chimps may be incredibly social animals, but they can also do an enormous amount of damage in a confined space, and they stripped the seven-acre enclosure down to a few tall trees in less than a decade.

When a neighboring farm came up for sale in 1995, Dave and I knew we had the solution. So we bought the thirteen-thousand-acre property, organized the Chimfunshi Wildlife Orphanage Trust, then donated the land to it. The only question then was how big to make the next enclosures. We considered every manner of configuration, including one grand scheme whereby a single fenced enclosure would be built of twenty-five hundred acres and all of our chimps would be released simultaneously in separate corners. That way, they could roam and stake out territory, much as chimps do in the wild. But the thought of all the fights that would take place—and the inevitable casualties—was too much to bear.

That's why two five-hundred-acre enclosures were finally opened on April 15, 2000, carved out of the forests and floodplains along the Upper Kafue River, with enough thick jungle and fruit groves and open grassland to allow the chimps to live how they'd want and where they'd want. Each enclosure was ringed by ten-foot electrical fencing and included the same concrete and steel handling cages we'd built at the smaller enclosures, but it was still the largest area ever set aside for captive primates; some estimates reckoned it was the combined size of ten zoos. It wasn't freedom—we knew that—but in a world where chimpanzees are hunted for meat and forests are decimated daily, it was probably as close to freedom as any of our chimps might ever get.

For Dave and myself, the huge enclosures represented the fulfillment of a dream, one we'd nurtured since it became apparent that raising orphaned chimpanzees was somehow more than just an accidental hobby for us. When people hear we have over eighty chimpanzees now at Chimfunshi, or see us struggling to pay all the bills, they always say, "Well, don't accept any more chimps and your problem will eventually take care of itself." But that's ridiculous. First of all, these chimps could easily live to be fifty or sixty years old, which is breathtaking when you consider that I'm seventy and Dave is seventy-three—our chimps will be alive long after we're gone. And what are we supposed to do when somebody tells us that if we don't take a chimp it will be euthanized? You can't say no then, can you? What if two baby chimps suddenly crop up somewhere in the Middle East, and then two more are found in a meat market somewhere in the Congo? I suppose some people would say, "Sorry, that's it, no more chimps," but not us. That's not our way. And so the chimps keep coming and coming.

But we decided to relocated only about half of our chimps into the new enclosures, since we just had too many babies and bachelors and outright misfits to take a chance on moving them all. It was agreed that Charley's family group of twenty individuals would be moved into one enclosure and that Chiquito's family group of twenty-five would go into the other. But when we began to discuss the logistics of such a move, we realized that no one had ever attempted to relocate that many chimps at once before. In the past, every time we'd moved chimps, we'd only moved a few at a time, carrying some in our arms, leading others by the hand, and drugging the bigger ones and shuttling them back and forth in a wheelbarrow. And the fact that all of our enclosures had been just a few hundred yards apart made things much easier.

This was more complicated. This was the relocation of dozens of chimpanzees, to a site that was an eight-mile drive along roads that were difficult at the best of times and almost nonexistent at others. If something were to go wrong—say a chimp got loose and ran off, or worse, attacked someone in the confusion—we could have a disaster on our hands. It was clear: Dave and I could not hope to move that many chimps by ourselves. It needed to be done properly or not at all.

Fortunately, we were blessed with good friends at the veterinary school at the University of Pretoria at Onderstepoort in South Africa, where Dr. Gerry Swan presides. He is an expert on animal treatment and care, and he literally wrote the book on animal anesthesiology that Dave and I refer to so often. When we contacted Dr. Swan and asked if he'd be willing to help, he agreed, offering to send a team, headed by his associate, Dr. Peter Buss, that would do all of the darting of the chimps and help monitor them as we moved them from their old enclosures to the new ones.

They had their ideas and I had mine, and when the vets announced that they were going to move fifteen or sixteen chimps a day, I nearly fainted. We knew nobody had ever tried to move this many chimps before, and I was terrified. I was adamant that the vets weren't going to do it in the way they said they were, which was to come up here and do it in five days. I said that if that was all the time they could spare, then I'd cancel it altogether. So they then arranged to come up for a longer period.

Peter, in retrospect, was brilliant, because he never said to me, "Right, this is how we're going to do it," or, "No, Sheila, you're wrong," or any of the things you might have expected him to say.

All Peter ever said was, "All right, Sheila, OK."

Not that he didn't keep trying. At one point, he said casually, "You know, we're quite happy to move fifteen chimps a day." And I'm afraid I exploded. "No way!" I snapped. "I want no accidents! This is not going to happen."

"All right, Sheila," he said. "OK."

And when the day came to begin darting the chimps and moving them, I had our staff set out moving crates, and I said, "Right, here's the boxes we're going to put the chimps in for the move." And Peter said, "Sheila, do you really think the boxes are necessary?"

"Look," I said. "We've got to drive them almost eight miles, and if anything happens on the road, we can't take chances."

"All right, Sheila," Peter said. "OK."

In retrospect, I can see that I had been upset for weeks before the move. My temper was getting shorter and shorter and I could feel myself getting tight in the chest. The pressure was immense, and I'm certain I was no joy to be around. I was just worried there was going to be an accident involving a chimp, and I knew I couldn't handle that. When Peter said they were going to move fifteen a day, I saw nothing but trouble; I was sure one of those moves would go badly.

Because these weren't just animals to Dave and me. They were precious individuals who came to us in a time of need and allowed us to help them, who had the nobility and strength of character to permit us to get close. I often say they forgave us, and I believe it's true. I know it sometimes seems as if tending to eighty chimpanzees is a lot of bother, but in reality, I get far more from the chimps then they do from me. I think I was like an overprotective mother, and some of the vets seemed more worried about me than about the chimps.

When the vets moved into position, the chimps knew right away that something was up. Their screaming and frantic attempts to escape were terrible to watch. We'd brought them all into the handling cages, and then the vets lined up outside the bars and shot them one by one with anesthetic darts. The poor chimps cowered and hid away, covering up as best they could. I told them that it was for their own good, they were going to wake up somewhere better, but it was no use. It was just heart-wrenching.

I think the one that hurt me most was Buffy, a fifteen-year-old female who'd been with us for only about eight months after spending several years alone and abused at an animal park in Zimbabwe. Her mate had been killed years before during an escape there, and she arrived with a deep distrust of white people, yet had made great strides at Chimfunshi. But she was in one of the central cages that day and when she saw the dart gun, she, of course, thought we were coming for her—even though she was one of the chimps we had no intention of darting and moving. But poor Buffy just kept thinking we wanted her and she was screaming and crying and running here and there. She'd obviously had a dart gun used on her many times before—maybe real guns too, for all I know—and she was petrified.

At one point, Peter asked me if I had any spare mattresses lying about, and it just so happened I had just recently purchased several dozen for the new Education Center we'd built, so we got them out and put them under the trees in the shade. Then, once the anesthetic took hold, the vets and our staff entered the chimps' cages and placed them on canvas slings with looped handles at each end. It took four men to lift some of our biggest male chimps, and the first thing they did after carrying the chimps out was weigh them, hooking the loops of the sling onto a scale that was strung from a tree. Then they eased the slings onto the mattresses, rolled the chimps off one by one, and gave them all complete medical examinations. They performed dental checkups, took temperatures, tested their hearts—the works. They also stitched up Tara, who'd suffered several cuts in a recent fight, and checked over all of the chimps' bumps and bruises.

There were nine vets in all—six from Onderstepoort and three whom we worked with regularly in Zambia—and they all came together. I was proud of our staff, too, the local Zambians who'd spent years working with these chimps, who'd learned to read their every gesture as they fed them and cleaned up after them day by day. Our guys were helping to turn the chimps over, moving them this way and that. They all worked so quickly and so professionally that it took your breath away. By this time, my notion of placing each chimp in a traveling cage had been discreetly ignored—and I could see the folly of insisting—so we got the seven-ton truck out and pulled up alongside the prone chimps, who were lying every which way on the mattresses. The mattresses were placed side by side on the flatbed of the truck, and after we'd got a full load, I climbed up with several of the vets and off we went—albeit slowly.

The new enclosures were situated just three and a half miles northwest of the Chimfunshi main compound, but that's three and a half miles as the crow flies—meaning, in central Zambia, during the dry season. But the floodplains were impassable at that time of year, so we were forced to go the long way around, via a bumpy dirt road that sliced this way and that through the forests on our land, a distance of almost eight miles. The vets and I sat on the back of the truck with the chimps, each vet carrying a syringe in his or her hand, and if a hairy arm or leg came up, the vet would rush over and jab the chimps again with knockout drops to keep them under.

At one stage, a female vet said, "I think this one's coming awake."

A male vet said, "Leave it a moment and let's see."

Suddenly, the male vet's leg went flying up in the air, and he was thrown over onto his back. Clearly, that chimp needed another dose.

We were a motley caravan, this dilapidated old red truck creeping over potholes and through the brush, followed by a long line of vans and trucks and cars carrying everything from well-wishers to television crews. Because we were forced to drive so slowly, the trip took close to an hour. And even though the road has always been terrible, it seemed particularly bad that day, and I thought we were going to get stuck a couple of times. But we didn't. The first journey was the longest of my life, for sure, and I felt every single bump in the road. I just couldn't get comfortable, and every time one of the chimps moved a head or a hand, I'd fear the worst, expecting them all to jump up and begin causing trouble.

But after that first trip, things got easier. We gained confidence as we went on, and after we'd moved the first few chimps, we started to get a rhythm. And once I realized how confident these vets were, I started to relax, too.

Although the official opening of the new enclosures was scheduled for April 15, we moved the chimps a few days ahead of time so they could get used to their new surroundings. We placed them in the handling cages adjacent to the enclosures at the new facilities and allowed them to sleep off the anesthetic, then fed them and kept them inside for a day or two so they could get adjusted. After they settled in, we even let them out for a brief stroll, just so they could get a sense of the new place and so we could see if there were any weaknesses—or "escape routes"—in the new enclosures, before a crowd was there to complicate maters.

Sure enough, our new enclosures were no match for Sandy. The fact that he was now the largest chimp at Chimfunshi—he weighed in at 165 pounds on the day of the move—also meant that he could do more damage than most when on the loose, so he'd been confined to a handling cage for many, many months. Yet we'd felt sure these new enclosures would hold him. After all, what chimp wouldn't love five hundred acres of forests and trees? But after being placed in the new enclosure, Sandy walked out of his cage, glanced at the chimps ambling off in the other direction, then hauled himself up and over the electrical fence and was gone. The electricity was working, too. But, as always, Sandy had already figured the odds, accepted the shock, and escaped.

Several chaotic hours later, we had Sandy back in hand after finding him in the forest seated on a log alongside one of our trustees, Stephan Louis, sharing a soda. There was nothing left to do but redart Sandy and return him to his cage over at the main compound. But on the way back, something suddenly went wrong. The small van Sandy was riding in screeched to a halt and everybody was shouting, "Stop! Stop! Stop!" and all at once the vets were rushing here and there as the call went up, "Has anybody got water?" It seems Sandy had begun to overheat in the back of that little van, even though he was unconscious, and his body temperature had gotten dangerously high. I think it was a combination of his having jumped the fence, which really got his adrenaline pumping, and the fact that the little van was rather stuffy—the vet riding with Sandy got heat sickness, too. But whereas the vet's body temperature went back down the minute he stepped out of the van, Sandy's just kept climbing higher and higher. The vets were very worried—you could see it in their faces—but nobody panicked. People with water bottles came forward and they put Sandy in this canvas sling and everybody started pouring water over him to cool him down. Then they laid him in the shade under a tree and everybody fanned him and massaged his arms and legs until his temperature came back down to normal and he was fine.

Looking back, I know that Sandy could very easily have died then, because I'm aware of what can happen under anesthetics. The body's functions of heating and cooling simply don't work as well under the drugs, and given how big Sandy is, he requires more anesthetic than most chimps to knock him out. And that was always my biggest concern with this move. But the vets never let on at the time, I think perhaps because I was there. They realized how stressful this was to me, and handled me with as much care as they did Sandy.

We finally got all of the chimps moved, and then came the official opening. We'd invited a large crowd to witness the event, and more than 150 people showed up, including most of our longtime supporters from places like Germany and Sweden and England. A number of the local dignitaries also attended with their wives and children, and a women's dance troupe performed. We had television crews from England and South Africa on hand, and Zambia's Permanent Secretary to the Copperbelt, Arthur M. Yoyo, presided over the affair and cut the ribbon to open the enclosures. My daughters spent days preparing sandwiches and snacks for the VIPs, and a long table groaned under all the food.

But when it came time to officially release the chimps, I was able to block out the crowds and the noise and to focus on just one chimp: Pal. It had been eighteen years since he'd come to us, a desperate little chimpanzee, his face torn open, his teeth smashed, and his body racked by dehydration and diarrhea. God, I'll never forget the smell; if death has a smell, that was it. Nobody thought Pal would survive, and on some of those lonely nights when I sat up nursing him or cuddling him through his nightmares, I had my doubts, too. Yet there he was, big and robust, with only those scars and that droopy lower lip to remind us of how sad he'd once looked.

My heart was in my throat as I placed a hand on the sliding metal door and peered into Pal's cage. I leaned in close. "I promised you this," I whispered. "Now off you go."

And with that, I pulled open the door and gave Pal his freedom. He stepped through the opening onto the sandy earth, followed closely by Tobar and Spencer, who immediately puffed themselves up to enormous sizes and began to display, waving sticks and dashing about as the crowd clapped and cheered. Before Pal joined in, however, he turned to look back at me, staring straight into my eyes. And maybe it was my imagination, but for just that magical second, I believe he was thanking me.

Before Pal arrived, Dave and I had believed our lives were winding down. But Pal and all the chimps that followed forced us to react, to do the best we could for them. Now there are chimpanzee sanctuaries all over Africa, and even a few in the United States and Europe, and some of them look to us for guidance. But I never quite know what to say. Dave and I just happened to be in the right place at the right time, with the right circumstances and the right amount of land. And we have had the good fortune of possessing the right sort of friends to back us. Our feeling is that if an animal comes to you in distress, you have to try and help it. And you can't give in to despair; you have to go on, no matter how hopeless the situation may seem. We've made mistakes from time to time at Chimfunshi, and there's certainly things I might do differently now if given the chance, but there's one thing I know for sure: The chimps have given me far more than I could ever repay.

Scientists say that genetically, chimps are 98.4 percent the same as humans, and maybe it's those similarities that are important—not the differences. We really aren't that far apart.

I think that's what Peter meant when he approached me near the end of the move, as I sat on the back of the seven-ton truck and watched the last chimps being offloaded. We were both exhausted. He and I had certainly had our differences at the start, but we'd come to see eye-to-eye as the relocation progressed, and I emerged as one of his biggest supporters. But now Peter looked almost apologetic.

I asked him what was wrong, and Peter just shook his head.

"Sheila, when I came here, I thought we were going to move a lot of chimpanzees," he said. "I didn't realize they were all part of your family."

Appendix

Chimfunshi Wildlife Orphanage Chimpanzees

Name	Sex	Year Arrived	Offspring / Birthdate
Pal	Male	1983	NA
Liza Do Little	Female	1984	Ingrid (F)–1991; Lori (F)–1995; Lionel (M)–2000
Girly	Female	1984	Goliath (M)–1991; Genny (F)–1997
Junior	Male	1984 (died 1986)	NA
Charley	Male	1984	NA
Bella	Female	1984 (died 1999)	Brenda (F)–1995
Spencer	Male	1985	NA
Tobar	Male	1985	NA
Boo Boo	Male	1985	NA
Donna	Female	1985	None
Chiquito	Male	1986	NA
Tara	Male	1986	NA
Cleo	Female	1986	Cooney (M)–1994 (died 2000); Colin (M)–1998
Rita	Female	1986	Renate (F)–1997
Sandy	Male	1986	NA
Cora	Female	1986 (died 2000)	Connie (F)–1996
Little Jane	Female	1987	Little Judy (F)–1995; L.J. (M)–2001

Name	Sex	Year Arrived	Offspring / Birthdate
Coco	Female	1987	Carol (F)–1996
Jimmy	Male	1988 (died 1989)	NA
Big Jane	Female	1989	Bijoux (F)–1992; Bob (M)–2001
Josephine	Female	1989	None
Grumps	Male	1990	NA
Mikey	Male	1990	NA
Sheena	Female	1990 (died 1990)	None
Pan	Male	1990	NA
Dora	Female	1990	Dolly (F)–1996
Goblin	Male	1990	NA
Choco	Male	1990	NA
Leben	Male	1990	NA
Milla	Female	1990	None
Pippa	Female	1990	None
Toby	Male	1990 (died 1994)	NA
Lucy	Female	1990 (died 1998)	None
Maggie	Female	1991	Miracle (F)–2000
George	Male	1991	NA
Misha	Female	1991	Max (M)–2001
Georgette	Female	1991 (died 1994)	None
Noel	Female	1991	Nikki (F)–1997
Diana	Female	1992	David (M)–2001
Zsabu	Male	1992	NA
Violet	Female	1992	None
Trixi	Female	1992	Tilly (F)–2001
Doc	Male	1992 (died 1998)	NA
Charles	Male	1994	NA
Tina	Female	1994	Thompson (M)–1995; Tess (F)–1998
Masya	Female	1994	None
Boogie	Male	1994 (died 1999)	None

Name	Sex	Year Arrived	Offspring / Birthdate
Tamtam	Female	1994	Thomas (M)–2001
Guenther	Male	1995	NA
Bobo	Male	1995	NA
Sampie	Male	1995	NA
Clement	Male	1995	NA
Brian	Male	1995	NA
Doreen	Female	1995	None
Barbie	Female	1997	None
Roxy	Female	1997	None
E.T.	Female	1997	None
Stephan	Male	1997	NA
Louise	Female	1997	None
Junior	Male	1999	NA
Alphie	Male	1999	NA
Buffy	Female	1999	None
Buddy	Male	1999	NA
Nick	Male	2000	NA
Bobby	Male	2000	NA
Eusebio	Male	2000	NA
Sonny	Male	2001	NA
Sinki	Male	2001	NA
Kambo	Female	2001	None
Val	Male	2001	NA
Kathy	Female	2001	None
Julie	Female	2001	None

Acknowledgments

Dave and I founded the Chimfunshi Wildlife Orphanage in 1983, but we owe the success of the sanctuary to so many others. With the arrival of chimpanzees came a succession of people who were dedicated before they came or who became dedicated after spending a little time with our extended family, and we are deeply in their debt.

Some of our earliest supporters remain so to this day, such as Dr. Shirley McGreal, Ingrid Regnell, Stephan Louis, and Richard Watson. Shirley's help is always forthcoming no matter who or what the trouble is, and both Ingrid and Stephan have been invaluable as fund-raisers and as friends. Richard is a chartered accountant who has looked after the purse strings without complaint since the day we started. Others, like Thomas and Margaret Cook, Eddie and Lil Brewer, Stella Marsden, Jack Stoneley, Julien Pettifer and Robin Brown, and Peggy and Simon Templar, offered encouragement and advice in the beginning, while Ken Pack, Karen Pack, Tamar Ron, and Sally Shiel were there when I needed them most.

We have three extra "daughters" in our extended family—Carole Noon, Nikki Ashley, and Brenda Santon—and each has woven herself into our lives forever by being the best friends a chimp could hope for. We also count Tess Lemon and Tamsin Constable among this group, even though those late evenings with a bottle of Irish Cream got a bit hair-raising.

Charles Mayhew is the founder and chairman of the Tusk Trust and one of our biggest supporters, and Steve and Carol Thompson were regular visitors to the farm before being transferred to South Africa. It was there that Steve founded the Friends of Chimfunshi, which continues to raise funds and awareness on our behalf. We have also received a steady supply of advice and support over the years from Karl and Katherine Ammann, Jacob and Ilse Mwanza, Sister Constancia Treppe, Dr. Jack Kyle, Madeline Gould and Herman Wadsworth, Duane and Lori Raab, Fred and Renate Winch, Professor John Jelos, Jenny Armstrong, Gerard de Nijs, Masanor "Mutsan" Hata, Harriet Crosby, Steve and Barbro Robinson, Roger and Deborah Fouts, Jackie Hanrahan, Martin Harvey, Karla Boyd, Linda Tellington Jones, Arlene Brahm, Julie Hart, Carol Rosenthal, Zoe Ball, Sarah Scarth, Susan Hales, Danie Van Der Walt, Patti Ragan, Kelli Murphy, Debbie Rogers, and the whole "50/50" TV crew, and the famed primatologist Geza Teleki, who once honored us by writing in a Christmas card, "It is a pity that some of us academics did not leave some things to you in the first place."

Then there is Jane Goodall, who first came to us in 1990 with Milla. Reading her books helped us so much during our early years as we struggled to understand chimps, but getting to meet and know Jane has been an inspiration in and of itself. What a wonderful, soft-spoken person she is—one who continues to do so much for our nearest relative.

The list of people who have braved terrible and dangerous conditions to rescue chimps on our behalf includes Clement Mwale, Doreen van Bierk, June Hicks, Tristan Davenport and Vicky Borman, Peter and Diana Ritter, Tim and Rosie Holmes, Bertie Roomer, Nick De Souza, Kathy Bader, Valerie Molyneaux, Dr. William George, Mike and Linda Garner, Sarah Christiansen, Elba Munoz, and Sven and Catrin Hammer. Organizations such as the World Society for the Protection of Animals (WSPA), the International Fund for Animal Welfare (IFAW), the International Primate Protection League (IPPL), the Great Ape Project (GAP), TRAFFIC, and the local SPCA have been among our strongest allies, as have Interair Airlines, British Airways and Voyagers Rentals.

I must also thank the wonderful team of veterinarians who gave of their knowledge and compassion to move our chimps in April 2000. Led by Dr. Peter Buss, they included Melvyn Quan, J. C. Botha, Liza Koster, Ina Kruger, and Ian Espy, and came with the support of Dr. Gerry Swan at the University of Pretoria at Onderstepoort.

Our family has given us strength and support throughout this endeavor; they include Lorraine and Ian Forbes, Diana and Pierre Fabel, Sylvia Jones and Don Bennett, Charles and Valarie Siddle, Tony and Linda Siddle, and all of our grandchildren.

This book would not have been possible without the dedication and determination of the wonderful people at Grove/Atlantic, Inc., particularly Morgan Entrekin, whose support was crucial. Brendan Cahill was a thoughtful and concise editor who suggested changes that always made the book better, and Michael Hornburg was an extremely diligent copy editor. I also must thank my agent, Nina Ryan, of the Cowles-Ryan Literary Agency, who believed in this project from the start. Her partner, Katherine Cowles, made valuable contributions, and Chandler Crawford handled the sale of foreign rights.

Thanks, too, to Quinn Atherton and Sandra Cress, who offered insightful suggestions as the book developed.

Finally, I would like to acknowledge the contributions of the chimps that are no longer with us: Junior, Bella, Cora, Cooney, Jimmy, Sheena, Toby, Lucy, Georgette, and Doc. May you all rest in peace . . .